TO SEE A WORLD

TO
SEE A
WORLD

JOHN W. HARRINGTON

Professor, Department of Geology,
Wofford College, Spartanburg, South Carolina

With 26 illustrations

Saint Louis

THE C. V. MOSBY COMPANY
1973

Printed in the United States of America
International Standard Book Number 0-8016-2058-9
Library of Congress Catalog Card Number 72-91625
Distributed in Great Britain by Henry Kimpton, London

To Casey

"To see a world in a grain of sand,
And a Heaven in a wildflower,
Hold Infinity in the palm of your hand,
And Eternity in an hour."

The Auguries of Innocence
WILLIAM BLAKE

BEFORE WE BEGIN...

This book is about understanding science, particularly the science of geology, and the way it is used to see our world. Understanding means to know enough about something so that every part of it is seen in proper relationship to all the other parts and to the whole. The following guide shows how the parts fit together. The literary couplets that preface each chapter are intended to reflect the meaning of the more technical material to follow. The frontispiece is a cartoon that represents the way the chapters, the parts, and the book serve as a platform from which *you* will be able "to see a world" once you have gained this level of understanding.

ACKNOWLEDGEMENTS

A small college is sometimes looked upon from the outside as analogous to a small automobile, cute but underpowered. This may be true in terms of certain types of research requiring extensive libraries and expensive laboratory equipment. However, in terms of rich conversations and interactions between students and colleagues of different specialties, the small college is a wellspring of inspiration. Creative achievement seems to be spurred by a critical mass of thinkers pushing and pulling one another along. In such an environment, individual departments are small and the critical mass must come from a coalition of disciplines. The members cannot share technical minutiae so they search for common grounds on which they all stand together. This makes the writing of a book like this one little more than an extension of coffee time. It also makes the task of giving credit for particular contributions a very difficult one. How can I remember which teacher or student first expressed a viewpoint that with a little twist has now become a part of me? I can only thank my colleagues en masse, with the hope that they savor their ideas as much on reading as they did in proclamation.

Specific credits must go first to my wife, who fought for all the revisions. Martha Wharton, lady d'Artagnan of all reference librarians, was never at a loss to find the facts. Dennis Dooley, master of the comma, served as sage and co-champion of the English language. This is a role he shared with Susan Dodge, an unusual combination of beautiful, indefatigable typist and born grammarian.

Critical review by William Romey, George White, Claude C. Albritton, Jr., Duncan S. Heron, Jr., and George Devries Klein helped immensely in restructuring the manuscript into its final form. Their wisdom is greatly admired. It is impossible to give full credit for their contributions. Finally, I wish to thank the students who kept their visits to my office at a minimum in order to allow me time to work throughout the summer. Student conversation is addictive to groping old professors.

JOHN W. HARRINGTON

GUIDE TO THE FLOW OF IDEAS

PART 1

Ideas on which all sciences are based

Chapter 1

Search men's governing principles, and consider the wise,
what they shun and what they cleave to.

Meditations IV, 38, Marcus Aurelius

The first principle of science

Joy is an emotion. There is a special kind of emotion that
sweeps over us when we form a new insight or completely understand some long-held information. We may grasp a joke heard
days ago and burst out with a gleeful chuckle at the most inappropriate moment. We may struggle fruitlessly with a problem
and then grasp the answer in an exhilarating flash. Robert Frost
called these intellectual joys the "feats and fruits of comprehension."

This book is about one kind of intellectual joy, the joy resulting
from geologic thinking. It has one purpose—to stimulate the comprehension that Blake meant in his line "to see a world."[1] The
purpose is grand enough but there must be a method to accom-
plish it. A familiar analogy is sometimes more easily grasped than
the still-mysterious project. It can then serve as a bridge into the
unknown. Ours comes from a first grader's preprimer. "Look,
Jane, look. See, Dick, see."[2] In spite of the words, children thrill
as they read them because they know they are actually reading.
Where Jane looks and what Dick sees have little to do with it.

In some ways this book is also a reading primer. The literature
of the earth is written in strange symbols made of minerals, rocks,
fossils, landforms, structures, and processes. Many of the stories are
hidden from most people because they cannot read the language.
Even subtitles in the form of college courses, textbooks, lectures
by guides in national parks, and a range of nature books are of
little help. The familiar earth is all too unfamiliar. Every language
consists of a vocabulary of words and a grammar that explains
how the words are strung together to give meaning. Oddly enough,   3

very little has been written about the grammar of the wonderful language of nature. Unlike the horrors of sixth-grade syntax, this is pure fun, for it is based on the joys of comprehension.

The place to start is with the single concept on which the entire discipline of science is founded—its first principle. From this base the entire structure takes form. First principles are rarely discussed, but like the axiom system on which plane geometry is based, the first principle determines all that follows. The discipline becomes more a result than a creation.

The first principle of science is the presupposition that nature *can* be understood. Heartened by this hope, scientists have taken on the tasks of searching out the state of nature as it exists. They are fortunate because there is something unique waiting for eventual discovery. Progress in research is marked by the addition of verifiable information. Each breakthrough is cumulative because some new aspect of the form of nature has been added to the still incompletely known picture.

This, too, is a remarkable facet of the philosophic problem of recognizing the state of nature. We must presuppose that there is an external reality that is revealed through the senses to any number of observers in just the same way. A new scientific breakthrough must pass this test, as it must be possible to repeat critical experiments and observe the same results each time.

The identification of DNA, the genetic coding device, is a good example. Watson and Crick did not create DNA out of an alphabet of letters or a Tinkertoy set, as Madison Avenue creates advertising copy out of a vocabulary of words. Their accomplishment was one of discovery. They found out how a fundamental facet of nature operates. It has been in operation for a billion years or so, controlling the emerging character of every new living cell. But not until 1953 did James O. Watson and Francis H. C. Crick make the discovery, aided by an almost naive disregard for the standard approaches through chemistry and genetics. Before this date the mechanisms of genetics were a mystery. Now we are much closer to knowing how they work. The discovery came about because two young men of diverse backgrounds had the brashness to work on such a problem. They accepted the presupposition that nature *can* be understood and boldly proceeded to prove it! They became Nobel laureates later—after their work had been verified by others.

The many accomplishments of modern science during the last few centuries provide ample evidence that the presupposition that nature can be understood is useful and that there is a unique state of nature to be discovered. It is difficult to discuss this aspect of science without sounding smug. There is room enough for humility in discussions of all the things that science has not yet discovered, but this does not belittle the truly glorious point that the presupposition is a sound one and that the target is real and can be defined.

The case can be made in a more convincing manner by dealing with common ideas rather than those viewed through the eyes of Nobel laureates. The following examples appear to be quite unrelated, yet each is simply an illustration that the behavior of a material is to some degree controlled by one aspect of the state of nature, its own internal shearing strength. In each case we shall make the common presupposition that the phenomenon can be understood; then we shall set about the task of understanding it.

Consider a small stream flowing to the sea some 200 miles away. If we ask why the flow occurs, the immediate answer is the pull of gravity on the water. Oh! But gravity is a force directed straight down toward the center of the earth, not laterally toward the sea many miles away. Comprehension requires breaking the problem into small units and seeing what is happening right before our eyes. Imagine a single cube of water, one foot in each dimension, standing as a block on the ground. It is difficult to imagine such a block standing for more than an instant without collapsing into a broad wet spot. We expect this to happen because the nature of water is common knowledge. Water has a low internal shearing strength. It has a molecular structure that partially bonds a dozen or so units together, but permits the slipping change of shape we call fluid flow. The old folk phrase "like molasses in January" brings to mind a fluid of higher viscosity, that is, higher internal shearing strength. It is much easier to imagine a cubic foot of molasses standing for several moments on a frigid January morning before it slowly bulges at the base and oozes out into a sticky sheet. Exactly the same change of shape would occur in the cubic foot of water. The only significant difference would be in the amount of time involved.

Simple multiplication tells us that there must be 1,056,000 of these water blocks set side by side in a stream that is 200 miles long. No one of them may collapse until the others downstream get out of the way. The entire stream is a great chain of collapsing water blocks whose rate of flow is controlled by the slope, riverbed shape (determining external friction), and the internal shearing strength of the water. With this sort of insight as a beginning, scientists have developed greater comprehension by measuring the rate of flow and describing the action in quantitative terms and rigorous detail. Even a novice scientist can experience a great sense of joy in standing beside a stream and watching it collapse as a series of small units on their way to the sea. Professor Grope, a typical bumbling clown of a fellow, loads all of his students on busses and transports them to a stream each year, just to show them this simple illustration of the first principle of science, the presupposition that nature can be understood.

In the second illustration we shall consider the characteristics of different types of shorelines and relate the visible forms to the internal shearing strength of the coastal rocks. Some coasts are bold and cliffed, as pictured in Felicia Hemans' poem *The Landing of the Pilgrim Fathers in New England:* "The breaking waves dashed high/on a stern and rockbound coast. . . ."[3] Other shorelines are flat and sloping, with sandy beaches that absorb the force of the waves, deflecting them upward to be halted by gravity rather than stopped abruptly by a sea wall. Professor Grope has written of this effect in his poem *"Carolina Coast in Winter."*

I stood, wretched in fear on the hard open sand,
    and watched the charge.
Balaklava could not have been too different,
    for the guns boomed,
And the great uniformed lines collapsed,
    as legs were cut from under,
Ending each new assault with swirling agonies.

Yet behind each broken rank rose another,
    from the gray green sea source.
Calmly forming, checking line, as if on parade,
    Then with heads tossed high,
Reared plumes spuming in the wind, the charge
    flung forward, flashing hooves and knees, until
The spent heroes fell at my feet in sweaty foam.

One rank rising higher than the rest,
    Stormed on in wild-eyed desperation,
In seething fury overran my position,
    Driving me back, wet to the knees,
A cold, craven, sopping shore-snark, who learned too late:
    Sloping sand is no final barricade.

The geology here is obvious. Loose grains of sand form a mass with a very low internal shearing strength. Wave action inbound on such a beach stirs the sand and yet can carve no long-lived cliff, for the backwash keeps the sand flattened at a low seaward slope. Cliffs a few feet high may be cut in dunes overnight during a storm or by an unusually high tide, but these too soon collapse from lack of internal strength. Geologists have ordered these concepts into rigorous statements of fact, but even the novice can savor the joy of knowing the rules under which the battle between land and sea is being waged.

The third illustration is an exercise in viewing the egoism of the human perspective through a play on the expression "terra firma." We expect certain things of the earth. We insist, for example, on a firm support beneath us. Here is man, walking on the outer edge of a stony ball 8,000 miles in diameter, kept alive by the energy from an exploding hydrogen "bomb" 93 million miles away. We know there are an almost infinite number of similar suns in the universe, but the one next to us is so far off that its light takes 4.29 years to arrive on earth. We do not know that there are any more habitable planets associated with the other stars, but it would be foolish to imagine that we are the sole reason for creation. In any event, we alone seem to tread the outer edge of our space vehicle, only partially mindful of the vastness of the universe. Even so, if the outer rim of the earth is unexpectedly a bit muddy and its internal shearing strength is reduced enough to threaten a shoeshine by as much as half an inch, we have been known to curse the heavens for imposing the inconvenience!

A different sort of illustration of the insight that may be gained through exploiting the first principle of science may be taken from the story that the moon tells about the mechanics of the creation of the solar system. Since there is no water or atmosphere on the moon, there is no common form of weathering, erosion, transportation, or deposition to destroy our window into the past. The little destruction that does occur is due to the pounding of falling meteorites. Moon rock is cratered, ground to dust, and scattered across the moonscape. Nonetheless the moon today is very nearly as it was at the time of its beginning(s). Some of the most obvious features on the moon are the great impact craters hundreds of miles across. We call them mares, or seas, for they were once believed to be true oceans. They are the relatively smooth, dust-covered surfaces on which the first moon landings were made in

1969 and 1970. Oddly enough, almost all of them are on the side of the moon facing the earth. There is only one very small mare-like feature on the far side. We did not know this until space rockets were orbited behind the moon to photograph the side that is never turned toward us. Enough is known now about these features to allow us to assume that they are gigantic impact craters. If they are impact craters, we may conclude that the mares were formed within a two-week period. The reasoning is interesting, if not too simple to be trusted.

The earth's gravitational force is much greater than that of the moon, and so an incoming flight of space pellets must have been caught in our gravitational field and directed toward the earth. The swarm must have been very concentrated, because the pellets that missed the earth and struck the moon (a nearby target only 30 earth diameters away) did so in less time than it took the moon to make half an orbital swing to expose its back side. Unless the speed of the moon has been altered greatly by the change in mass during the impact, this time was less than two weeks, for the moon makes one revolution around the earth every 27⅓ days. If any more time had been available and the incoming pellet swarm more spread out, the back side of the moon would have been pounded and made to look like the side that faces earth.

It is most likely that the earth as well as the moon would have been battered during this short period of mare formation. Therefore the next question is the date of the event. Studies of the mare floors[4] indicate that a very complex history has been recorded in the layers of dust and minor craters. The upper few centimeters of dust that have been sampled by the astronauts are about 3.5 billion years old. Analysis of the characteristics of the minor craters suggests that portions of the mare floors have changed little in the course of the last 4 billion years. Since the mare basins must be older than the partially rearranged dust that fills them, the original depressions are very old indeed.

Putting it all together via the first principle of science, we can say that a cataclysmic event damaged both the earth and the moon during a two-week period that occurred much more than 3.5 billion years ago! Interestingly enough, geologists have been puzzled for years over the fact that we have never been able to find rocks on the earth that are more than 3.3 billion years old. The space program has supplied the answer. In the early days of the solar system a major infall of meteorites reconstructed the face of the earth. It

must have been quite a show. We are fortunate to have missed it
and yet be able to enjoy it in this vicarious manner.

The key to discovery is curiosity. Geology is an outdoor science, and this requires a special sort of energetic curiosity, first to go and see and then to interpret the collected data. Dr. Edwin H. Colbert, the ever-young curator of vertebrate paleontology at the American Museum of Natural History in New York, was 65 years of age when he led an expedition to the interior of Antarctica to demonstrate the significance of dinosaur bones found at Coalsack Bluff. Here is his first report to a colleague, Dr. Schaeffer, at the museum.

Dear Bobb: We did it! On our first day of field work we found a cliff full of Triassic reptile bones. . . . This happened yesterday. The bones are not articulated—it is a stream channel deposit. But they are numerous and in good condition. Should give a varied fauna. We are getting prepared for an intensive collecting program. The locality is Coalsack Bluff. Only a few miles from our camp. The cliff [a low series of small cliffs] faces north into the sun, and is thus relatively warm. We are sending a release to the N. S. F. [National Science Foundation]. . . . The N. S. F. has first rights, since this is an N. S. F. project done primarily with N. S. F. funds. . . . We are tremendously excited. This really pins down continental drift, in my opinion. Antarctica had to be connected in the Trias. [signed] A very excited old man.[5]

The phrase "continental drift" may require some explanation. Alfred Wegener, a German paleoclimatologist, made the first serious scientific proposal that the continents we know today as separate units were joined in larger groups about 200 million years ago. The concept of separation became known as continental drift. Wegener and his followers compiled a tremendous amount of data in support of the idea, but general acceptance was withheld until a mechanism to move not mountains but whole continents was hypothesized and accepted. A portion of the next chapter is devoted to this controversy and to the way in which a gradually growing enlightenment is reflected in textbooks of the times.

Wegener was no armchair geologist. He made four expeditions to the Greenland Ice Cap and perished there at the age of 50 years while attempting to cross a portion of it. The idea of continental drift came to him in 1910 while he was looking at a map of the world. The edges of the continents seemed to fit together perfectly if the intervening oceans were disregarded. Like Colbert and the bear who went over the mountain, Wegener was curious, and he had "to see what he could see." Wegener searched Greenland for the bedrock evidence of a link between the continents of Europe and North America. Colbert was excited because he knew that certain types of terrestrial dinosaurs were reptiles that could not have been present on the Antarctic continent unless it had been connected to the other major land masses where the same species were found. Our point is that discovery requires a remarkable combination of unbounded animal energy and an inquiring mind.

These qualities have existed through time. Pliny the Elder made the first recorded field check of the eruption of a stratovolcano. On August 24, 79 A.D., near Naples, Italy, Mount Vesuvius began the eruption that was to bury Herculaneum and Pompeii in beds of ash and mud. The account was set down by Pliny the Younger, the nephew of the famous naturalist. Pliny the Elder was in command of the Roman fleet at Misenum, where his attention was drawn to this scene.

A cloud, from which mountain was uncertain, at this distance, was ascending, the appearance of which I cannot give you a more exact description than by likening it to that of a pine tree, for it shot up to a great height in the form of a very tall trunk, which spread itself out at the top into a sort of branches; occasioned, I imagine, either by a sudden gust of air that impelled it, the force of which decreased as it advanced upwards, or the cloud itself being pressed back again by its own weight, expanded in the manner I have mentioned; it appeared sometimes bright and sometimes dark and spotted, according as it was more or less impregnated with earth and cinders. This phenomenon seemed to a man of such learning and research as my uncle extraordinary and worth further looking into. . . . Hastening then to the place from which others fled with the utmost terror, he steered his course direct to the point of danger, and with so much calmness and presence of mind as to be able to dictate his observations upon the motion and all the phenomena of that dreadful scene. He was now so close to the mountain that the cinders, which grew more abundant and hotter the nearer he approached, fell into the ships, together with pumice stones and black pieces of burning rock: they were in danger too not only of being aground by the sudden retreat of the sea, but also from the vast fragments which rolled down from the mountain, and obstructed all the shore. Here he stopped to consider whether he should turn back as the pilot advised him, "Fortune," he said, "favors the brave; steer to where Pomponianus is."[6]

Pomponianus was then at Stabiae, only a few miles along the coast from Pompeii and much too close to Vesuvius for safety. Apparently Pliny the Elder overexerted himself and suffered a heart attack,[7] for he died quietly the next day, a pillow tied upon his head for protection against the falling stones.

The eruption of Vesuvius in 79 A.D. is the oldest volcanic activity of which a detailed description is available. There have been a number of other eruptions duplicating the pattern since that time. Pliny's account is still useful, for good observations are timeless parts of the record of science.

A final illustration of the use of the first principle of science shows something of the cumulative way in which understanding is gained by the addition of ideas from many people over the years. In this next case the observations were made on the earth's gravitational field, and understanding grew as new light was shed on the foundations of mountain systems. The study is very important and still lies on the frontier of geologic research. We will only look at the first accomplishments, which proved that the roots of mountain ranges were different from the more uniform parts of the earth.

More than 250 years ago accurate measurements of the gravity field were made by counting thousands of swings of a pendulum and then using these to calculate the precise time required for one swing. Pendulum behavior had been understood since the classic works of Galileo (1564-1642) and Newton (1642-1727), when the expression linking the strength of the earth's gravitational field with the pendulum became a standard tool of physics. In 1735 the French physicist Bouguer was interested in determining the shape of the earth and the amount of flattening at the poles due to the expansion of the equatorial radius by rotational forces. In the tradition of all field men who must go and find out for themselves, Bouguer made pendulum studies in Lapland and on the west coast of South America near the equator. One of his measurements was taken near sea level and another high in the Andes Mountains. Bouguer was surprised to find that the effect of the "extra mass" of the mountains below him was much less than he had anticipated.

A full century later the British surveyor Sir George Everest, for whom the highest mountain the world is named, undertook the task of making a triangulation net for northern India. A net of this type is composed of a series of interlocking triangles that have been surveyed so precisely that error is reduced by multiple cross-checks of distances and angles. Such nets are used to fix base points for more detailed surveys of local areas. As an extra precaution, Everest measured latitude positions in two different ways. One used the old system dating from about 225 B.C. Measured angles between a fixed star and different points on the earth vary according to latitude because the points occupy different positions on a sphere. Each angle is measured in a vertical plane between the horizontal reference and a line to the star. In the second system, Everest measured horizontal distance from the southern to the northern base point and then converted this arc length to a change of latitude. In the 375-mile-long traverse there was a difference between the results of the two systems amounting to a 500-foot uncertainty in the latitude of the northern point near the foothills of the Himalayas. The work was so precise that the discrepancy, equivalent to the length of less than two football fields in the distance between New York and Montreal, was too much! The figures were rechecked and the difference remained. Everest realized that this was not a simple error. Instead there was some basic factor in the systems that kept them from producing identical results. He knew that this factor was the effect of the mountains on the spirit-level base lines used to orient the instruments. Within 20 years Dr. G. B. Airy, the British Royal Astronomer, and J. H. Pratt, the Archbishop of India, had developed separate solutions to the distribution of densities in the rocks beneath the mountains to account for the measured effects. The details of their models have been improved during the last century, so that we now have a fairly good picture of the hidden mountain roots. Our concerns are now those of the ever-intriguing question "Why?"

Ideas may be understood best if they can be viewed against a broad background rather than in isolation. Perhaps the first principle of science should be viewed in the context of some of mankind's other intellectual activities. Although it is rarely considered, such disciplines as education, law, government, and religion also operate on first principles that are really just simple presuppositions.

The first principle of education is the presupposition that learning is possible. The catch here is that learning begets knowledge and that knowledge can be divided into two main categories. One is rote knowledge. It is imitative in nature and stems from the process of "monkey see, monkey do!" The other is cognitive knowledge, obtained through creative personal insights. George Bernard Shaw distinguishes between the two very sharply in a dialogue between Cleopatra and an old musician and teacher in his play *Caesar and Cleopatra.*

Cleopatra: I want to learn to play the harp with my own hands. Caesar loves music. Can you teach me?

Musician: Assuredly I and no one else can teach the Queen. Have I not discovered the lost method of the ancient Egyptians, who could make a pyramid tremble by touching a bass string? All the other teachers are quacks: I have exposed them repeatedly.

Cleopatra: Good; you shall teach me. How long will it take?

Musician: Not very long: only four years. Your majesty must first become proficient in the philosophy of Pythagoras.

Cleopatra: Has she (indicating a slave girl who has been playing) become proficient in the philosophy of Pythagoras?

Musician: Oh, she is but a slave. She learns as a dog learns.

Cleopatra: Well, then, I will learn as a dog learns; for she plays better than you. You shall give me a lesson every day for a fortnight. After that, whenever I strike a false note you shall be flogged; and if I strike so many that there is not time to flog you, you shall be thrown into the Nile to feed the crocodiles. Give the girl a piece of gold; and send them away.

Musician: But true art will not be thus forced. . . .[8]

Science is a search for insight. Knowledge accumulates as a result. The old musician was right; one must first have some grasp of basic ideas before new concepts can be meaningful. Unfortunately educators do not agree on the relative values of rote and cognitive knowledge. There is a cleavage between teaching scholars that does not exist between scientists engaged in their researches. Although some teachers are satisfied with a "monkey do" performance, the best demand thoughtful understanding. This may be defined as knowing enough about something to be able to see every part of it in proper relationship to all the other parts and to the whole. The first principle of education is best applied when both students and teachers equate learning with understanding.

The first principle of law is the presupposition that society operates best under a fixed set of rules. These come from three sources: the authority of custom, legislative statute, and prior judicial decision. By and large at least two of these sources are conservative in nature, tending to hold a society on the course of the past. There is no built-in homing device toward an ultimate dream of justice, and there is no unanimity among cultures as to what is the best or only set of rules. It is not surprising that some degree of inconsistency in the interpretation and application of laws has been observed through the ages.

The first principle of government is the presupposition that the most desirable form is that which produces the most good for all of the people. The words "one," "produces," "most good," and "all" are meaty but subject to various interpretations, so that governments also have been known to become a bit unstable at times. This is particularly true if the presupposition is twisted to mean the most good for the most people. Minorities have been known to reject this with revolutionary results—1776, for example!

There are two great types of religions: the theistic, such as Christianity, and the atheistic, such as Buddhism. Their first principles are quite different. Theistic religions are based on the presupposition that there is a mutual concern between God and man. When the religion is broadened so that it becomes an effort to define these concerns, the system is then parallel to that of science. This similarity to science diminishes in cases of the more doctrinaire religions, for science is not concerned with deductive reasoning within a totally closed system. Atheistic religions are based on the presupposition that man has an ultimate concern. This is an open-ended statement and leaves room for many interpretations. Tillich has used the term "ultimate concern" to imply unconditional seriousness on the part of every involved individual.

Any tendency that scientists may have toward a callow pride in their absolute possession of the method for arriving at the truth may be dispelled by reviewing scientific history. It is a labyrinth of false starts, blind alleys, hedgehopping discoveries, fortunate accidents, brilliant insights, and crumbled doctrines. Blind alleys such as alchemy occupied protochemists for a century or two, but alchemy led eventually to the field of scientific chemistry. How much easier the birth of science would have been if research had been directed toward the nature of air instead of the coincidence between the colors of base and precious metals! It was the discovery of oxygen that put research on an open-ended track, not the transmutation of copper to gold or lead to silver.

What is the final limit of scientific research? Nothing has yet been discovered to indicate that nature is too complex to be comprehended by a composite of all human intelligence. Perhaps it is not meant to be so, but then there is a built-in limitation to the act of viewing. We are unable to comprehend that which cannot be conceptualized. Ancient man saw stars as bright spots in the night sky, but did not conceive of them as suns radiating energy in compensation for a loss of mass as hydrogen is converted to helium. We do not know at what level man's ability to conceptualize will fall short of visualizing the realities of nature. It may be that we are in the final golden age of science, an age in which the tools and intelligence of man are functioning at nearly maximum capacity. At present the view of science shared by most practitioners is that of an ever-expanding, ever-successful system. It may be a naive viewpoint, but it will serve until we discover our limitations.

ANNOTATED REFERENCES

1. Blake, William. 1888. The poems of William Blake. Walter Scott Publishers, London. 282 pp. (This poem is found on p. 208.)

2. Robinson, Helen M., Monroe, Marion, and Artley, A. Sterl. 1950. Sally, Dick and Jane. Scott, Foresman & Co., Glenview, Ill. 48 pp. (This is not a direct quotation but it is an example of the characteristics of the old-fashioned preprimer.)

3. Hemans, .Felicia. The landing of the pilgrim fathers in New England. In Felleman, Hazel (ed.). 1936. Best loved poems of the American people. Garden City Publishers, Inc., Garden City, N. Y. 670 pp.

4. Ronca, L. B. 1971. Ages of lunar mare surfaces. Geol. Soc. Am. Bull. **82:**1743-1748.

5. Lear, John. 1970. The bones on Coalsack Bluff, a story of drifting continents. Sat. Rev., Feb. 7. pp. 47-51. (This is indeed an adventure story. Dr. Colbert had written a book on the achievements of outstanding paleontologists. Then, as the date of his retirement drew near, he made this discovery, which ranks with any in paleontology.)

6. Adams, Frank Dawson. 1938. The birth and development of the geological sciences. Dover Publications, Inc., New York. 506 pp. (The account of Pliny the Younger is found on pp. 46-47.)

7. Zirkle, Conway. 1967. The death Gaius Plinius Secundus (23-79 A.D.). Isis **58:**553-559.

8. Shaw, George Bernard. 1906. Three plays for puritans. Brentano's Publications, New York. 301 pp. (The quotation about rote and cognitive learning is taken from Act IV, pp. 161-162. There are a number of other scenes in this play that are also relevant. Caesar's comments on the burning of the library at Alexandria are a reflection of Shaw's opinion of the military mind. In Act II (p. 115) an exchange between Caesar and Ptolemy's tutor, Theodotus, is a biting comment on teachers who pretend to be able to instruct men of action and decision:

   Caesar (turning to Theodotus): And you, sir, are . . .?

   Theodotus: Theodotus, the King's tutor.

   Caesar: You teach men to be kings, Theodotus. That is very clever of you!

Chapter 2

Ah! What avails the classic bent
   And what the cultured word,
Against the undoctored incident
   That actually occurred?

            The Benefactors, Rudyard Kipling

The fabric of science and the crisis of contradiction

The previous chapter established the philosophic base on which all science rests. This one shows how the work is carried on. We shall take our reasoning method from Sherlock Holmes and expand it into that used by scientists. This is not as odd as it may seem, for geologists and detectives both work with effects and attempt to establish the causes. A reprise of Holmes in action should set the mood. This classic example is taken from the *Study in Scarlet.* This is their first adventure together. Dr. Watson has not known Holmes long and is a little doubtful that the man really has the genius to see where others are blind. The scene opens at 221B Baker Street, London. Holmes and Watson are looking out the window. Watson is speaking.

"I wonder what that fellow is looking for?" I asked, pointing to a stalwart, plainly dressed individual who was walking slowly down the other side of the street, looking anxiously at the numbers. He had a large blue envelope in his hand, and was evidently the bearer of a message.

"You mean that retired sergeant of Marines," said Sherlock Holmes.

"Brag and bounce!" thought I to myself. "He knows that I cannot verify his guess."

16   The thought had hardly passed through my mind when the man

whom we were watching caught sight of the number on our door, and ran rapidly across the roadway. We heard a loud knock, a deep voice below, and heavy steps ascending the stair.

"For Mr. Sherlock Holmes," he said, stepping into the room and handing my friend the letter.

Here was an opportunity for taking the conceit out of him. He little thought of this when he made that random shot.

"May I ask, my lad," I said blandly, "what your trade may be?"

"Commissionnaire, Sir," he said gruffly. "Uniform away for repairs."

"And you were?" I asked with a slightly malicious glance at my companion.

"A sergeant, sir, Royal Marine Light Infantry, sir. No answer? Right, sir." He clicked his heels together, raised his hand in salute, and was gone. . . .

"How in the world did you deduce that?" I asked. . . .

"It was easier to know it than to explain why I know it. If you were asked to prove that two and two made four, you might have some difficulty, and yet be quite sure of the fact. Even across the street I could see a great, blue anchor tatooed on the back of the fellow's hand. That smacked of the sea. He had a military carriage, however, and the regulation sidewhiskers. There we have the marine. He was a man of some amount of self-importance and a certain air of command. You must have observed the way in which he held his head and swung his cane. A steady, respectable middle-aged man, too, on the face of him—all facts which led me to believe that he had been a sergeant."[1]

Part of the charm of the Sherlock Holmes stories lies in the way Doyle startled his readers by first stating conclusions and then having them proved correct. Holmes always produced enough supporting evidence to verify his guesses. The tie between evidence and conclusions depended on the principle of "least astonishment" in a relationship that may be diagrammed in this way:

$$\text{Facts } 1 + 2 + 3 + N \longrightarrow \left(\begin{array}{c}\text{Imply via least}\\\text{astonishment}\end{array}\right) \longrightarrow \text{Conclusion(s)}$$

The system of analysis is very old. Aesop used it in his fables 2,500 years ago. The story of the blind men and the elephant is typical. Each of the blind men felt a different part of the elephant and concluded that the whole beast was like his one sample. There must be some coordination of experience before the correct conclusion (that is, the whole elephant) may be assembled from the parts through the principle of least astonishment. Lacking this coordination, the blind men postulated a wide range of untenable hypothetical forms.

Aesop intended the fable to show that we are all blind at different levels, depending upon the experience we have had and can use to produce rational conclusions. The system that works so well in fiction is almost identical to the one scientists use. The chief difference is that scientists attempt to define their facts in absolute terms and recognize least astonishment solutions against a background of shared technical rather than common human experience. The use of technical experience that is beyond the understanding of laymen has given science a reputation for practicing mumbo jumbo. The only means of clarifying this is to begin at the beginning and assemble the entire structure again in scientific terms.

There are three parts in the expression: facts, reasoning, and conclusions. Philosophers consider a fact to be an "objective datum of experience." The curious Professor Grope was not sure exactly what was meant by this and asked Sergeant Sanders, a Pinkerton detective, to give another definition. The sergeant said, "A fact is a sure thing!" Grope then asked a philosopher with a mathematical bent what a sure thing was and received the answer, "A sure thing is an event with a statistical occurrence probability of one." Grope took this back to Sergeant Sanders to find out what was meant by "a statistical occurrence probability of one" and was told, "You can bet on it!" From this Grope has adopted the scientific definition of a fact as "a sure thing you can bet on." According to this definition, the observation of Aesop's blind men that the aspects of an elephant were like a rope, a tree, a snake, the side of a house, a large piece of leather, and so on were not facts at all. It is no scientific surprise that these did not total an elephant.

Philosophers use other names for the principle of least astonishment; among them are "logical simplicity," "parsimony," and "Occam's razor." If we accept as a definition of simplicity "the avoidance of unduly complicating hypotheses," we can see why S. F. Baker feels that "the notion of simplicity has [a] certain attractiveness, yet at the same time [it is] repelling in its vagueness."[2] How then can a successful venture like modern science be based on this seemingly vague principle? The answer lies in the way in which the principle is used.

Scientists consider each new conclusion to be something less than fact. A new conclusion is taken as a prediction about the possible (and hopefully probable) singular state of nature. Then it is tested to see whether it can be judged true or false. If after exhaustive research the conclusion cannot be disproved, it is used as fact and becomes part of a larger pattern. The system is not unlike the working method of the sculptor, who releases a form from the enclosing marble by chipping away the stone that is not needed. The idea that a straight line is the shortest distance between two points is a least astonishment conclusion. Although this axiom cannot be proved true, no other line between two points is as short as the straight one. Therefore the practical approach is to postulate that this is the case and proceed with the great Greek game of plane geometry.

Philosophers in classroom discussions are likely to underplay the value of the principle of least astonishment by their choice of illustrations. One could point to a particular chair and then make a claim that the chair was probably in the room ten minutes or so before class time. If the chair is bolted to the floor, the students would feel reasonably sure that this is the probable situation. However, if the chair is of the simple, straight-backed variety, the students would immediately suspect that the philosopher was making some sneaky point. Perhaps he had taken the chair out of the room and then put it back just to show how vague the principle could be. He would not be so cavalier with the possibility that the Washington Monument was moved and put back ten minutes before class time.

Laymen sometimes imagine that science operates through a system of laws that nature has to follow. Ohm's law, Boyle's law, Hooke's law, and the law of gravity are some of the familiar examples that come to mind. Odd as it may seem, these, too, are just least astonishment conclusions stating that under a particular set of circumstances certain relationships have held true for all known cases. Therefore the scientist is justified in assuming that these same relationships may be expected to hold true for all future cases. Any deviation would be shockingly unexpected. When Einstein proposed a relativistic mechanics for relationships at velocities approaching the speed of light, he found scientists incredulous. Then the ideas were tested and found to be true, augmenting the Newtonian mechanics that had stood for over two centuries.

Geology is an historical science. Many of the natural experiments are studied hundreds of millions of years after they occurred. Their analysis requires a special type of least astonishment reasoning called the principle of uniformity (uniformitarianism).[3] This reasoning tool, as emphasized by James Hutton and John Playfair at the end of the eighteenth century, put geology on a sound scientific basis.

The principle of uniformity is a least astonishment comparison between geologic processes actively producing observable results today and geologic features of some antiquity. The basic presupposition of the principle is that ancient processes differed very little in form, rate, and scale from those we see in action. Many illustrations of the use of this principle occur throughout the book. Look for them. You will discover that the principle was used, at least by implication, by Herodotus in the fifth century B.C. and by Chinese philosophers in the twelfth century A.D.

The greatest limitation of the principle of least astonishment is that it imposes a conservative bias on creative thinking. Conclusions that vary too greatly from the expected view of reality are automatically tuned out until all other ideas that are apparently more reasonable are tested and found wanting. The saving grace for science is the singular state of nature toward which all ideas that are proved correct must converge. Scientific literature plays an important role in preserving ideas that may have been shunned for all the wrong reasons and yet are fundamentally correct. Research is conducted in much the same way a pack of North Carolina blue tick hounds flushes a fox. A trail is abandoned as soon as it is known to be false, and the pack circles about for the scent again.

Since the entire fabric of science is woven on the principle of least astonishment, we must be sure to understand it before we proceed. Therefore one more illustration will be presented, even at the risk of intellectual overkill. When we board a commercial airliner, we have every right to expect that the wings will support the plane, the engines will drive it forward, and the pilots will know how to make it fly. Specialized scientific and engineering expertise justifies these expectations. However, when the Wright brothers began their studies of flight, none of these expectations were foregone conclusions. Their methods of investigation began with an idea and ended with the age of flight, testing each intermediate least astonishment prediction and conclusion in turn. First they assumed that powered flight on wings was possible. They built a wind tunnel using a power-driven fan and cardboard mailing tubes to make the air flow in a laminar rather than in a turbulent manner. Later a better wind tunnel of larger size and more reliable characteristics was constructed. They found the mono-wing the most efficient lifting device, but one that could not be built strong enough for their purposes. Two wings, one above the other, could be made as a truss with tension and compression members to yield great strength with very little weight. The wind-tunnel experiments led to the design of a full-scale model capable of supporting a man in flight.

21
*The fabric of science
and the crisis of
contradiction*

The next step was to learn to control an airplane. The Wright brothers needed something bigger than the wind tunnel, so they wrote to the United States Weather Bureau to determine where the winds were strong, reliable, and smoothly flowing. Kitty Hawk, North Carolina, was suggested because the winds flowing across the Atlantic and onto the continent were strong and dependable. Furthermore, the soft sands of the Kill Devil dunes would minimize injuries to plane and pilot. There the Wright brothers flew wings as tethered kites until the control systems were perfected. The problem of a power plant that could create its own wind in flight was solved by building a gasoline engine. Finally, on December 17, 1903, the miracle of Kitty Hawk came off on schedule. Although the Wright brothers had announced their plans, only six observers came out to watch as the cumulative result of dreams, assumptions, and least astonishment predictions became a conclusion.

This example is not the same as that of the learned philosopher arguing about whether or not a chair has been moved. The fabric of science is constructed on an interlocking series of predictions and tests. Each part of it is at first a naive subjective guess, but only until it can be tested!

Fig. 1 is a schematic diagram of the fabric of science as it is assembled by its practitioners from their most basic facts. Implied relationships between the facts constitute the first level of conclusions. Once these are proved, they too become facts supporting a second level. The fabric of interwoven ideas is expanded to some level "N," where our present inability to answer questions marks the frontiers of knowledge. Beyond is the still incompletely known state of nature—"that which *is*"—toward which the search for understanding is directed.

During this golden age of science there are many hundreds of thousands of well-trained and well-paid practitioners around the world engaged in searching, testing, and predicting. The older distinctions between physics, chemistry, and biology have begun

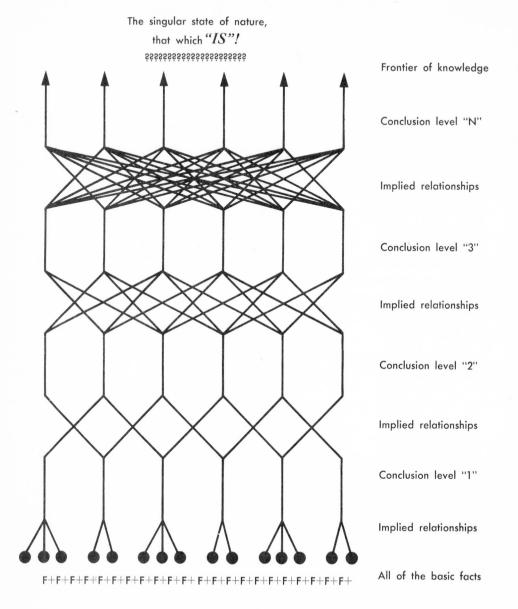

The singular state of nature, that which *"IS"!*

???????????????????????

Frontier of knowledge

Conclusion level "N"

Implied relationships

Conclusion level "3"

Implied relationships

Conclusion level "2"

Implied relationships

Conclusion level "1"

Implied relationships

All of the basic facts

F+F+F+F+F+F+F+F+F+F+F+ F+F+F+F+F+F+F+F+F+F+F+

Fig. 1. The fabric of science. This device illustrates the logical pattern of the fabric, woven of the implied relationships between facts and conclusions. The ultimate level of achievement is the discovery of that which "is."

to fade. At one time the physicists claimed the atom from the outside in, and the chemists from the inside out. Obviously it belongs to both and to neither, for it also belongs to the biologists and to the full range of scientists working in other disciplines that get down to the atomic level. Astronomers studying atomic reactions in the stars are as concerned with the atom as are the physicists. Biophysics and molecular biology have linked what were once separate fields. Geology involves all branches of science through geophysics, geochemistry, paleontology, astrogeology, planetology, oceanography, and meteorology. Classical geography has been a quasi-social science since the time of Herodotus (484?-425 B.C.). It is best to view the fabric of science as an unbroken spectrum from the full range of the social sciences including sociology, geography, government, and economics, through the life sciences including biology, to the plexus of the physical sciences. Each part has its frontiers, but these are shared with neighbors. Gains in one area yield clues and tools for advances in other areas. There are no renaissance men who command the entire field, but there are many who appreciate the fabric and its growth.

Perhaps the fabric will have more meaning if a single idea, for example, the law of gravity, is traced from its earliest postulation to the modern frontier. At least 200,000 years ago Neanderthal man must have thrown rocks as weapons. He probably did not pose any philosophic questions about why rocks fell back to the earth, but the matter of increasing the range by throwing upward at an angle of about 30 degrees must have been standard practice. The influence of Aristotle (384-322 B.C.) may have unduly delayed experimental study. His belief that falling bodies traveled at velocities proportional to their masses was not disproved until Galileo (1564-1642) made the famous pendulum experiments we have already mentioned. Newton (1642-1727) discovered that there was a proportionality in the force between two masses that depended upon their product divided by the square of the distance between them. In 1798, nearly a century later, Cavendish (1731-1810) managed to evaluate the universal gravitational constant and formulate the force-mass expression we use today. We now have a reasonably good picture of how the force of gravity operates, but the frontier question of why there is an attraction between separated masses is one of the mysteries to be answered by future research. Thus we see that the fabric of science is open-ended after 200,000 years of experience.

The words "hypothesis" and "theory" are used a good deal in scientific discussions. How do they fit into the fabric? An hypothesis is a conjecture or guess about some unproven relationship. It falls under the operation of the principle of least astonishment. Ideas remain hypothetical until they can be tested. Research, competition between ideas, and least astonishment solutions form checks and balances to eliminate poor hypotheses and replace them with conclusions. A theory is an intermediate stage in the testing procedure, marking a level of understanding that from all appearances is probably valid. With more evidence, a theory is upgraded to the status of fact. There is no court or tribunal that makes this decision. The words are simply used by a practitioner of science to express his judgment or opinion on appropriate status. One man may call the concept of human evolution "factual," and another may refer to the "theory" of evolution. There are no true scientific controversies despite the common use of the terms. Things that are controversial are not understood well enough to permit any viewpoint to be treated adamantly. Uncertainty is healthy and serves as a spur to creative testing. The apparent controversies that have been held in the name of science have been discussions about data, hypotheses, and predictions.

Science cannot exist in a state of suspended activity, for its very life is based on activity. The search for better ideas and the abandonment of poor ones *is science.* If this process were to cease, the body of accumulated knowledge would drift toward the realm of dogma. Any question of enforcing a moratorium on scientific progress is as ridiculous as the old nursery rhyme:

Mother, Mother, may I go swimming?
   Yes, my darling daughter.
Hang your clothes on a hickory limb,   .
   But don't go near the water.

Grant, then, that science builds its fabric through change. How does this take place? Are scientists a group of masochists who bear the pain of their world in dissolution with some abnormal delight? Hardly—scientists are human. They accept change because it is forced on them by two different kinds of pressure. One is internal. Scientists have agreed that they will accept the most logical patterns of thought in the tradition of the legal dictum "the truth, the whole truth, and nothing but the truth." Once a scientist

25

*The fabric of science
and the crisis of
contradiction*

knows that some facet of his work is not true, he is forced to admit it and go on. This is internal pressure; it is very effective. What if he does not do this and just fudges a little? The interlocking nature of investigation is such that he will be caught and either tuned out by his colleagues or the whole thing charged off to human error. Fudging will not produce a scientific result that is compatible with the singular state of nature. This is a dead end that is not pursued.

The second pressure is brought to bear by anomalous data. Anomalies are deviations from expectations. Scientists have agreed to accept all real data and deal with it. Once a bit of anomalous data is confirmed, it is capable of disrupting the old fabric and forcing the creation of a new one. The operation takes place in this way. First there is the happy state of ignorance masquerading as knowledge before the discovery of the anomaly.

$$\text{Facts } 1 + 2 + 3 + N \longrightarrow \left(\begin{array}{c}\text{Imply via least}\\\text{astonishment}\end{array}\right) \longrightarrow \text{Conclusion(s)}$$

Now add the anomalous information and the expression becomes intolerable.

$$\text{Facts } 1 + 2 + 3 + N + \text{Fact(s) anomalous} \longrightarrow \text{Imply ???}$$

The anomalous facts create the *"crisis of contradiction!"*

$$\text{Fact(s) anomalous} \longrightarrow \text{Contradict} \longleftarrow \text{Previous conclusion(s)}$$

Thomas S. Kuhn, in his classic study *"The Structure of Scientific Revolutions,"* has examined the case histories of many such crises of contradiction and worked out the historical patterns that are followed. We shall fit the discussion together in his own words by excerpting a number of short quotations from their proper context and sequence. Some of the verb tenses have been changed to give better cohesiveness. This method of presentation may be most unconventional, but it does communicate a remarkable set of ideas in the author's own style. The process Kuhn discloses is that of uprooting the happy state of normal science and creating, by revolution and new insight, an entirely different fabric within which normal science proceeds again.

Kuhn considers the scientific paradigm to be the unit of the fabric that changes.

... [new paradigms] implicitly define the legitimate problems and methods of a research field for succeeding generations of practitioners. They [new paradigms] are able to do so because they share two essential characteristics. Their achievement is sufficiently unprecedented to attract an enduring group of adherents away from competing modes of scientific activity. Simultaneously it [a new paradigm] is sufficiently open-ended to leave all sorts of problems for the redefined group of practitioners to resolve.

Discovery commences with the awareness of anomaly, i.e., with the recognition that nature has somehow violated the paradigm-induced expectations that govern normal science. It then continues with a more or less extended exploration of the area of the anomaly. And it closes only when the paradigm theory has been adjusted so that the anomalous has become the expected. Assimilating a new sort of fact demands more than additive adjustment of theory, and until that adjustment is completed—until the scientist has learned to see nature in a different way—the new fact is not quite a scientific fact at all.[4]

In our own terms, the fabric of science has been restructured, so that the familiar facts-conclusion(s) relationship is now:

$$\text{Facts } 1 + 2 + 3 + \text{ Facts New } + \text{ N} \longrightarrow \left(\begin{array}{c} \text{Imply via a new} \\ \text{capacity for least} \\ \text{astonishment} \end{array} \right) \longrightarrow \text{Conclusion(s)}$$

Kuhn points out:

Almost always the men who achieve these fundamental inventions of a new paradigm have been either very young or very new to the field whose paradigm they change. And perhaps that point need not have been made explicit, for obviously these are the men who, being little committed by prior practice to the traditional rules of normal science, are particularly likely to see those rules no longer define a playable game and to conceive another set to replace them.[4]

The operation of normal science with the new paradigm continues until there is another crisis of contradiction. Examples on the grand scale may be taken from Newton's mechanics, Maxwell's electromagnetic work, Einstein's relativity, the chemistry of Priestly, Lavosier, and Dalton, and most recently the work of Watson and Crick on DNA. Nobel prizes are not given in the field of geology, so that the revolutionary work of Alfred Wegener (1880-1930) is not as well appreciated as some other great insights are.

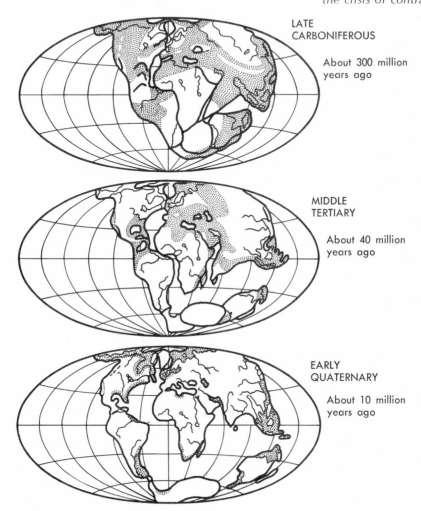

LATE
CARBONIFEROUS

About 300 million
years ago

MIDDLE
TERTIARY

About 40 million
years ago

EARLY
QUATERNARY

About 10 million
years ago

Fig. 2. Wegener's conceptualization of continental drift during the course of the last 300 million years. Stippled areas on these maps mark positions of shallow seas that partially covered the continental blocks from time to time. (From Wegener, Alfred. 1924. The origin of continents and oceans. Translated from the third German edition by J. G. A. Skerl. Methuen & Co., Ltd., London. 212 pp. Fig. 1, p. 6, has been used as a base for this illustration by permission of the copyright holder.)

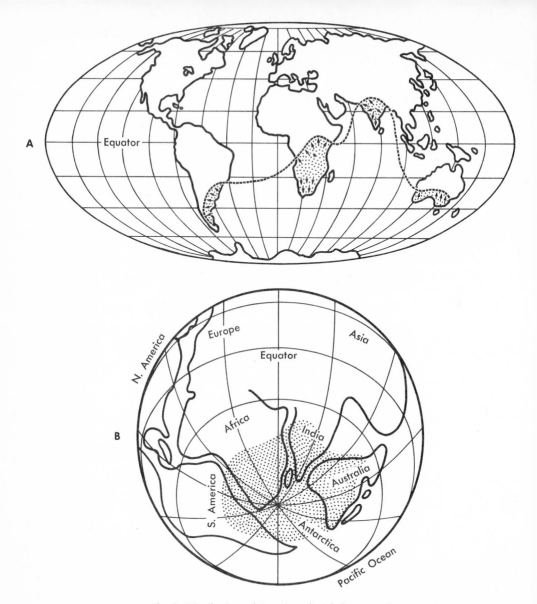

Fig. 3. Distribution of Permian glacial deposits, the anomaly that triggered a scientific revolution. **A,** Areas covered by continental glaciers in the Permian period, about 250 million years ago. The arrows show the direction of ice flow. **B,** Areas glaciated in the Permian period reassembled at the South Pole, as interpreted by Wegener in 1924. (After Holmes, Arthur. 1965. Principles of physical geology, ed. 2. The Ronald Press Co., New York. 1288 pp. Figs. 539 and 540 used as bases for Fig. 3 by permission of the copyright holder.)

Yet more than half a century ago he too structured a revolution, the fruits of which are just now being savored.

The Wegenerian revolution is an interesting one to follow, for it was not adopted by geologists without a good deal of conservative foot dragging. A healthy climate of scientific skepticism still remains in some quarters, but most geologists now accept some form of continental displacement as factual. Fig. 2 is a reproduction of Wegener's map of the protocontinent he called Pangea and the way he believed it had split into the present continental blocks. This theory was first published in German in 1915. An English translation of the third German edition was published in 1924.[5] Scientists soon began to realize that the widespread continental glaciation that occurred in the Permian period some 250 million years ago was a major anomaly that had to be faced. Fig. 3, *A*, is a map of glacier distribution showing the directions of ice movement. The position of glacial materials in torrid India just north of the equator simply did not fit the preconceived ideas of the permanence of continents and ocean basins. If, as in Fig. 3, *B*, the continents are reassembled in the Wegenerian way but placed so that the center of this ice sheet is at the south pole, the anomaly is explained. However, the explanation would require the equally difficult maneuvering of continents. The paradox hinged on evidence of glaciation that could not be denied, yet scientists were unable to accept continental movement by forces that could not be explained.

When a standard textbook first appears on the market, it is a compilation of the knowledge or fabric of a discipline. Research that improves understanding forces textbook authors to revise their work every few years just to keep pace. Old textbooks are really history books, and as such they are quite useful. It is interesting to trace the way in which successive editions of a single textbook treat the problem of continental drift over five decades. The chosen case is ideal, for although there were co-authors over the years, their professional lives overlapped and the books form an unbroken chain of simple revisions. The first edition of *A Textbook of Geology,* by Louis V. Pirsson and Charles Schuchert, was published in 1915 and predates any widespread appreciation of the concept of drift. The data that stirred Wegener are treated in the same way that fiction writers have treated Atlantis. Blocks of the crust may move up and down but not sideways as Wegener postulated. The last revision, published in 1969, was written to take into account the revolutionary concept that by this time was generally accepted.

Five years after Wegener first published his theory of continental drift, Professor Schuchert wrote the following discussion in the second revised edition of his own textbook. One of his areas of technical specialty was in the construction of paleogeographic maps, so his concepts of the distribution of land and sea through time are a particularly good choice to represent the fabric of science as it was known in 1920. Under the heading "Permanency of Continents and Ocean Basins," the following excerpts give the background.

Older and Newer Views.—In the earlier days of geology it was held that there was no stability in the continents and oceans as such, and that there had been complete interchange between them. Even Sir Charles Lyell taught that all parts of the ocean bottoms had been land. Now, however, most geologists hold with Dana [another famous Yale professor, 1813-1895] that the ocean basins and the continents have in the main, although not in detail, been permanent features of the lithosphere at least since the close of Proterozoic time [about 600 million years ago]. There is likewise much agreement among geologists in the belief that the ocean basins are sinking areas, also spoken of as the "negative areas" of the lithosphere, because the sum of their crustal movements is downward; and in general it appears that the oceanic basins have gradually attained not only greater depth but somewhat enlarged area as well. On the other hand the continents are the rising masses of the lithosphere in relation to sea-level, and for this reason are also called the "positive areas," because the sum of their movements is upward. . . . *In general, however, it may be said that the present oceans and continents have been more or less permanent features, and that they have been where they are now, with moderate changes in their outlines, since their origin in Proterozoic time or earlier* . . . [italics mine].

Proof of Permanency.—The proof that there has been no complete interchange between the continents and the oceans is seen in the following facts: 1) On the continents there are almost no deposits of deep sea origin. . . . 2) The marine deposits on the continents are nearly always those of shallow seas, in fact, not at all unlike those now accumulating on the continental shelves . . . . 3) It is now established that the lithosphere is denser and therefore much heavier beneath the ocean basins than under the lands, making it impossible for them to have interchanged their position without destroying the equilibrium of the outer shell.[6]

Professor Schuchert is about to catch himself in his own trap. He has just made the case for the permanence of continents and ocean basins. He will now make the case against permanence, almost as if he does not realize it. The crisis of contradiction is often expressed so subtly that the practitioners of science fail to notice it!

Examples of Continental Fragmenting.—As an example may be cited Madagascar . . . . No naturalist doubts its former connection with Africa because of their similar animals, and yet the channel of Mozambique which now separates it from the mainland is from 240 to 600 miles wide . . . . To the northeast of Madagascar lie many small islands, the Seychelles, and to the northwest the Comoro group, all of which are also held to have been parts of Africa and Madagascar. Not only this, but many biologists and geologists hold that all of these lands are but parts of the comparatively recent land Lemuria . . . .

Gondwana Land.— . . . Besides the facts given above, there is much other evidence of a geologic, paleontologic, and zoologic character relating to the distribution of the plants and animals since the Paleozoic [about 230 million years ago], tending to show that Brazil was once widely connected with northwestern Africa across what is now the deep Atlantic Ocean. This lost continent is the "Gondwana Land" [from a district of the same name in India] of Neymayr (1883) and Suess (1885), and of the zoogeographers, a vast traverse of land stretching from the northern half of South America, across the Atlantic to Africa and thence across the Indian Ocean to peninsular India, including Lemuria. It was in existence throughout the Paleozoic [geologic time between 600 and 230 million years ago], but the Atlantic bridge and Lemuria sank into the oceans during the Mesozoic. Gondwana when complete was comparable to another transverse land of the north, "Eria" or Holarctica, which existed when North America was continuous with Greenland and Eurasia across Iceland to the British Isles.[6]

We must stop here for a comment to be sure that these ideas are placed in a context that can be followed. Schuchert has been caught in a paradox. The continental masses share so many factors in common that they must have been connected, but he still must insist that the ocean basins have always been dense, heavy, low blocks. His solution is typical of an effort to include all the data. In effect he says that both things happened; the continents were connected by land bridges that did not exist because the ocean basins were always low blocks of more dense materials. We can see how Wegener, the young meteorologist and exactly the sort of man Kuhn identifies as the iconoclastic type, could begin to look for a more defensible hypothesis.

Schuchert considers the glacial climate of the Permian period [280 to 230 million years ago] as an undeniable phenomenon that presents a most interesting pattern of distribution.

> For nearly fifty years [prior to 1920] geologists have described unmistakable glacial deposits of Permian age in the continents of the southern hemisphere. . . . The Permian glacial formations are found on either side of the equator from about 20° to 35° north and south latitudes. . . . The evidence is now unmistakable that early in Permian times all of the lands of the southern hemisphere were under the influence of a glacial climate as severe as the polar one of recent times. . . . What brought about this great change in the climate of Permian time, and why it was apparently restricted to the southern hemisphere are as yet unsolved problems.[6]

Schuchert used the words "apparently restricted" to avoid having to face the glacial deposits of India north of the equator, even though they are carefully identified in an illustration in his text. The summary of the paradox is given in two sentences under a continued discussion of Gondwana Land.

> Belief in the existence of Gondwana is wide-spread among European geologists, but many American workers do not yet believe in it, mainly because they hold strong to the theory of the permanence of the oceanic basins and continents. Without this continent, on the other hand, paleontologists cannot explain the known distribution of Permian life, and, further, its presence is equally necessary for the interpretation of the peculiar distribution of marine faunas beginning certainly with the Devonian [405 to 350 million years ago] and ending in the Cretaceous [135 to 65 million years ago].[6]

We must add that the concept of continental drift is now known to explain these points; the continents were split before the Cretaceous period, and the gap was filled by new ocean floors of dense lavas.

The 1941 edition of the textbook was co-authored by another great Yale professor, Carl O. Dunbar. The prologue to this edition contains two points that show that the basic ideas of the authors had not changed greatly with respect to the permanence of continents and ocean basins.

The oceans have always been where we see them now.... The floods (seas spread across the continents) are as often withdrawn by the recurrent deepening of the oceans, but *there has never been a general interchange in position between continental masses and the basins of the oceans* [italics by Schuchert and Dunbar].[7]

The authors had good reason for this viewpoint. This is just what the geologic data imply, except that the splitting of supercontinental masses and formation of new oceans between is not considered seriously. Yet a much softer attitude toward the rigorous view of continental stability is seen in the section dealing with Permian glaciation.

The most remarkable feature of the Permian glaciation is the *distribution* [italics by Schuchert and Dunbar] of the ice sheets. They were chiefly in the southern land masses and in regions which now lie within 20° to 35° of the equator. This circumstance more than any other has lent attractiveness to the belief in "continental drift." If the southern continents were united to Antarctica until after Permian time, the glaciation may not have spread into low latitudes. A later "drift" of these continents towards the north would account far more easily than any other means yet postulated for the distribution of glacial deposits. But this premise itself is still in the realm of speculation![7]

Two decades later the crisis of contradiction was about to erupt into a full-blown revolution. By this time the senior author, Professor Schuchert, had died, and Dr. Dunbar, carrying on alone, put out another edition of the textbook, this time entitled *Historical Geology*.[8] The rigid ideas about the permanence of continents and ocean basins are missing, but the comments on Permian glaciation and continental drift remain unchanged. Dr. Dunbar no longer attempts to straddle the paradox without taking sides, but he is not yet able to accept the hypothesis of continental drift. In the next few years geophysical data that prove a form of continental displacement will overwhelm the last resistance, but in 1960 the old bastions still held.

The 1969 edition, co-authored by Dr. Dunbar and a third Yale professor, Dr. Karl M. Waage,[9] caught up with the times. New paradigms—continental drift, sea-floor spreading, plate tectonics, and the new global tectonics—were competing for adoption. A new chapter, entitled "The Restless Crust," was added to replace the abandoned ideas about the permanence of continents and ocean basins. The authors emphasized the spirit of change by heading the chapter with a verse from Tennyson.

There rolls the deep where grew the tree,
   O Earth what changes hast thou seen!
There where the long street roars hath been
   The stillness of the central sea.

Although this particular scientific revolution is not over, the corpses have been kicked aside, and activity is now directed toward building a new paradigm under which a new normal science may progress. In the words of Kuhn, "the scientist has learned to see nature in a different way." By this we mean that the once strange and unacceptable ideas have now become those that are least astonishing. Lateral displacement of continents in space have become so much a part of the new thinking that any suggestion of fixed positions would be considered by most geologists as too astonishing for belief.

Astonishing and least astonishing are relative terms. This entry in *The Diary of Samuel Pepys* for May 23, 1661, is a case in point.

To the Rhenish wine house, and there came Jonas Moore, the mathematician to us; and there he did by discourse make us fully believe that England and France were once the same continent, by very good arguments, and spoke of many things, not so much to prove the Scripture false as that the time therein is not well understood. . . .[10]

The seeds of the anomaly are old indeed, but they did not sprout and bear fruit for some 300 years! Science has a halting, struggling history. Viewed in retrospect, it is not the sacrosanct gem its practitioners would like to imagine.

1. Doyle, A. Conan. Study in scarlet. In Haycroft, Howard (ed.). 1892. The boys' Sherlock Holmes. Harper & Row, Publishers, New York. 336 pp. (This quotation is taken from pp. 20-21. Doyle uses the identification of the ex-soldier by bearing and suntan in a number of his stories. The case of the Greek interpreter is most notable, for both Sherlock and his brother Mycroft get into the act.)

2. Barker, S. F. 1957. Introduction and hypothesis: a study of the logic of confirmation. Cornell University Press, Ithaca, N. Y. 203 pp. (The quotation about the attractiveness of simplicity is taken from p. 162. Used by permission of Cornell University Press.)

3. Simpson, George Gaylord. 1963. James Hutton and the philosophy of geology. In Albritton, Claude C., Jr. (ed.). The fabric of geology. Addison-Wesley Publishing Co., Reading, Mass. 372 pp.

4. Kuhn, Thomas S. 1962. Structure of scientific revolutions. In International encyclopedia of unified science. Vol. 2, No. 2. University of Chicago Press, Chicago. 172 pp. (The quotation on the definition of new paradigms is taken from p. 10; the information on the pattern of discovery from pp. 52-53; and the discussion on the characteristics of individuals who achieve breakthroughs from pp. 89-90. Used by permission of the University of Chicago Press.)

5. Wegener, Alfred L. 1924. The origin of continents and oceans. Translated from the third German edition by J. G. A. Skerl. Methuen & Co., Ltd. London.

6. Pirrson, Louis V., and Schuchert, Charles. 1920. A textbook of geology (ed. 2, revised). John Wiley & Sons, Inc., New York. 1026 pp. (The quotations on the permanency of continents and ocean basins were taken from Part II, "Historical Geology," pp. 463-466. Schuchert's comments on the glacial climate of the Permian period were taken from pp. 758-760. His summary of the paradox of continental glacial deposits in India and on both sides of the equator is taken from p. 761. Used by permission of John Wiley & Sons, Inc.)

7. Schuchert, Charles, and Dunbar, Carl O. 1941. A textbook of geology (ed. 4). John Wiley & Sons, Inc., New York. 544 pp. (The material quoted from the prologue is found on p. 2; the discussion of Permian continental glaciation is found on p. 291 of Part II, "Historical Geology." Used by permission of John Wiley & Sons, Inc.)

8. Dunbar, Carl O. 1962. Historical geology (ed. 2). John Wiley & Sons, Inc., New York, 500 pp.

9. Dunbar, Carl O., and Waage, Karl M. 1969. Historical geology (ed. 3). John Wiley & Sons, Inc., New York. 556 pp. (This reference and verse are found on p. 69 ff. Used by permission of John Wiley & Sons, Inc.)

10. Morshead, O. F. (ed.). 1926. Everybody's Pepys, the dairy of Samuel Pepys, 1660-1669. Harcourt Brace Jovanovich, Inc., New York. 615 pp. (Pepys' entry for May 23, 1661, is found on pp. 93-94. Used by permission of the copyright holder.)

PART II

Application of the ideas on which all sciences are based to the problem of individual learning

Chapter 3

There is nothing worse than thinking you
are thinking and not knowing what you
are thinking about.

Mista Bob Cates

Student learning and the
crisis of contradiction

Once Professor Grope appreciated how scientists discover the
unknown through the crisis of contradiction, he began to wonder
to what degree other learning is produced in the same manner. A
dog is housebroken through contradictions. A child learns through
painful contradiction not to step into a bathtub full of water without first checking the temperature. What actually takes place in
the child's mind after he has learned this lesson may be a clue
to appreciating the role of the crisis of contradiction in higher
education. Imagine that we are watching a child approach his bath.
He steps up to the tub, pauses, extends his toe to test the temperature of the water, and then reacts. During this pause he must be
thinking. At one level the thought might be: "Don't forget what
happened last night. Wow!" If the water had been too hot the
evening before, the experience would have provided negative rein-
forcement from an external source. The child has learned as a dog
learns. At another level the child might have developed an internal
judgment system on which to base his actions. The thought process could be something like this: "If I knew how hot this water
were, I would know what to do about it. Let's collect the data and
make a logical decision." Such a child might be a little unreal,
but he would certainly be a philosopher.

*Application of the
ideas to the problem of
individual learning*

Most student learning occurs on some level between these two extremes. When motivation is external, students learn as dogs learn. Pressure from many external sources produces some learning and much of the poor scholarship to which teachers are accustomed. Pressures to get a grade, to please parents, to get a degree as a union card, or to join a fraternity are like whips that raise welts for failure. They are negative in nature. On the other hand, any student who has a high degree of internal motivation based on personal judgments of the material is a functioning practitioner of the discipline. Linus Pauling received his first Nobel Prize for research begun while he was still a college sophomore. Such motivation could not be instilled in anyone by negative methods.

There is little difficulty in educating a thinker, for he learns by extending his own world view. The crisis of contradiction functions for him in the same manner as for professional scientists. Even though such a student may not have much information or many skills at his command, he has already become a part of the scientific community. He has adopted its way of life. He knows that he knows only those things that can be related through the checks and balances of logic. The rest he holds in abeyance for testing. He will give up a view that cannot be defended and adopt one that has been proved correct.

Charles Darwin at the age of 22 years is a classic example of this type of learning behavior. Semantics of personality are very interesting. The name Charles Darwin brings to mind a grizzled 50-year-old man beset by the quasi-religious controversies that stormed around him after the publication of the *Origin of Species* in 1859. The mature practitioner was real enough, but he learned his trade as all of us do, bit by bit. Fortunately the record of how he learned it is clear enough to serve as a pattern for others to use.

By many standards, young Charles seems to have been an indifferent student. He entered the University of Edinburgh at the age of 16 years to study medicine but soon gave that up as a bad job. Then he began to prepare for the ministry at Cambridge. This was not to his liking either. In January, 1831, he was still undecided about his major interests and enthusiasms. His life seemed to be without passion or pattern. He still had two semesters of residence required at the University before he would be eligible for a degree to be taken without honors. The change came over him abruptly. Sir Gavin de Beer tells the story in his scientific biography *Charles Darwin*.

In the summer of 1831 Darwin accompanied Adam Sedgwick on a geological excursion in North Wales, when an incident occurred that made a deep impression on him. Darwin had learnt that a shell of the tropical mollusk "Voluta" had been found in a gravel pit near Shrewsbury, and he expected that Sedgwick would be delighted at "so wonderful a fact as a tropical shell being found near the surface in England." To Darwin's astonishment Sedgwick said that if the shell really had been embedded in the gravel pit [an anomalous situation] "it would be the greatest misfortune to geology, as it would overthrow all we know about the superficial deposits of the midland counties" of England. The gravel pit was [dug in] a glacial deposit, and nothing before had made Darwin realize that "science consists in grouping facts so that general laws or conclusions may be drawn from them," a lesson that he put to good use.[1]

In this rare account the young Charles Darwin seems to have made his own formulation of the expression:

$$\text{Facts } 1 + 2 + 3 + N \; \Big| \!\!\longrightarrow \binom{\text{Imply via least}}{\text{astonishment}} \longrightarrow \text{Conclusion(s)}$$

Furthermore, Darwin discovered that the fabric of science was constructed of parts linked to one another by a rational progression of thought. He had already found that learning languages was extremely difficult for him. Therefore it seems reasonable to imagine that up to this time Darwin did not know there was any way to learn except through rote memorization.

To be intellectually honest we must correct de Beer on the matter of identifying the Shrewsbury gravels as glacial at this time. The concept of continental glaciation was accepted for northern Europe by Louis Agassiz in 1837 and not extended to the British Isles until 1840.[2] Even then, Sedgwick tended to favor the idea that the superficial deposits and scarred bedrock surfaces beneath them were due to icebergs floating in an overlapping sea, abrading the land and dropping gravels that had been frozen into the ice.[3] The contrast of floating ice and the tropical Voluta shell would be startling enough to explain Sedgwick's reaction. Apparently he felt that Darwin had been handed a hoax.

Sedwick was one of the great field geologists of all time. Darwin could not have had a better teacher. The very important recognition of the significance of an anomaly and the potential crisis of contradiction was clear to both men. On the face of it, Sedgwick

42

Application of the
ideas to the problem of
individual learning

has been cast as a scientist afraid of an anomaly, and Darwin cast as an apprentice disappointed at having an anomaly discredited. Voluta is the villain that cannot be contained in the mold of least astonishment.

The year 1831 was a great one for Darwin. In August he received an invitation to accompany 26-year-old Captain Fitzroy on H. M. S. *Beagle* as a "naturalist without pay" for a five-year, 40,000-mile voyage of discovery. One of the professors at Cambridge had recommended him for the position with the amusing note, "I have stated that I consider you the best qualified person I know of who is likely to undertake such a situation. . . .[1]" The *Beagle* sailed on December 27. Darwin's description of the cramped quarters he shared with two ship's officers and a large table could have been written by any college freshmen about his own environment. "I have just room to turn around and that is all."[1] Such were the surroundings of the body, but Darwin is not famous today because he was able to accept privation while working for five years without pay. The surroundings of the mind were much more important.

Charles Darwin's future gift to the world was to be a cohesive account of the changes in living things through geologic time. The accuracy of the account depended on his ability to relate the fossil record and the events these data represented in proper chronologic fashion. It was fortunate that this voyage began in 1831, for the intellectual tools for deciphering the rock record were not available a few decades earlier. One evidence of Darwin's good fortune may be found in his library on board ship. He sailed with Volume I of Lyell's *Principles of Geology*, which had been published in 1830. Volume II of the three-volume set was waiting for him at the port of Montevideo, Uruguay, the following October. These books were the first essentially modern texts ever written on geology. They contain a remarkably complete fabric, tying facts together with an interwoven logic of least astonishment solutions. Darwin was therefore able to read the literature of geology in the language in which he would find it written throughout the world, the language of minerals, rocks, fossils, landforms, structures, and processes. He was fully aware of a vast span of geologic time and of the manner in which the parts could be ordered to see the whole. He was even aware of the concept of the progressive development of organic life and of the way in which it was linked to the study of fossils "in situ."

A case in point may be taken from the topic headings for Chapter 9 in the Lyell book that Darwin read while on board the *Beagle*: "Theory of the progressive development of organic life considered," "Evidence in its support wholly inconclusive," "Vertebrated animals in the oldest strata," "Differences between the organic remains of successive formations," "Remarks on the comparatively modern origin of the human race," "The popular doctrine of successive development not confirmed by the admission that man is of modern origin," "In what manner the change in the system caused by the introduction of man affects the assumption of uniformity of the past and future course of physical events."[4]

When this young man sailed away on his great adventure, his destiny was predicated on these four factors: (1) He had the tools with which to work! (2) He knew he had the tools! (3) He had the whole world as his laboratory! (4) He knew that too! Perhaps his genius is best seen at this level, rather than in terms of the totality of his life's work. As each journey begins with a single step, so did Darwin's productivity begin with intellectual awareness. A sense of destiny and unbounded animal energy seem to distinguish great men. Newton paced his study for hours on end, and Boyle moved his barometer with the zest of a child at play.

One other intellectual factor must be reemphasized, for it certainly contributed to the success of the voyage and to Darwin's achievements afterward. He seems to have been alert to the significance of anomalous data as the key to a new world view. His experience with the tropical shell may have taught him to search for unexpected variations in the pattern of otherwise normal facts. Nothing would seem to be more commonplace than to find ground finches among the birds inhabiting the Galapagos Islands. This archipelago is centered on the equator about 650 miles west of Ecuador. Darwin visited the islands for three weeks in September and October of 1835. In those days, naturalists were unsure of the meaning of variations within species and tended to think of a species in rather rigid terms. Darwin noted that the finches on the different islands in the group were distinguished by the shapes of their bills. He did not know what to make of this data, but sensing an anomalous situation, he placed this note in his journal: "I have stated, that in the 13 species of ground finches, a nearly perfect gradation may be traced, from a beak extraordinarily thick, to one so fine that it may be compared to that of a warbler."[1]

44
*Application of the
ideas to the problem of
individual learning*

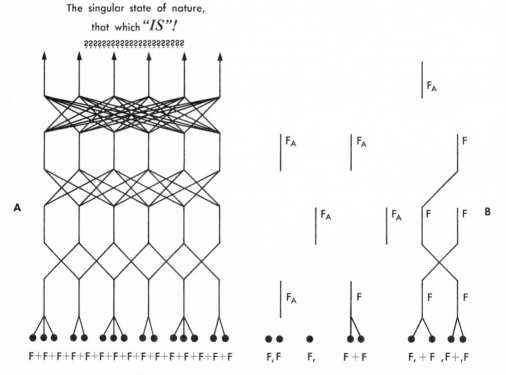

Fig. 4. Comparison of two views of the fabric of science. **A,** The fabric as seen by practicing scientists. **B,** The fabric of isolated scientific facts seen by students.

The significance of this observation lay unnoted for some time, even after Darwin's return to England. By 1845, a full decade later, another note appears in the second edition of the *Journal of Researches of the Voyage of the Beagle:*

Seeing this gradation and diversity of structure in one small, intimately related group of birds, one might really fancy that from an original paucity of birds in this archipelago, one species had been taken and modified for different ends.[1]

The theory (fact) of evolution is so well accepted today that we seldom consider the logical ground on which it stands. Prior to Darwin's publication of the *Origin of Species*, naturalists were bound by the preconception that species had been created as distinct and separate entities, much like the atoms of Greek physics. Ideas about the transmutation of species by changes in forms through time were not defensible due to the lack of an explanation of the method of change. In these two citations from the 1835 field notes and the 1845 revisions, we can see the ideas upon which Darwin constructed the hypothesis of natural selection as a mechanism for change beginning to take shape.

All of this information about Charles Darwin and the technicalities of his work has been introduced to make a single point that is very important to the discussion of student learning. The young Charles Darwin learned in exactly the same manner as the mature practitioner he was to become. He built a fabric of science and altered it to comply with anomalous data as changes were forced upon him. The world has not been the same since Darwin saw variations in the bills of a few ground finches and heretically explained them as modifications of a parent stock.

Students today can learn in the same way. Fig. 4 is a diagrammatic view of the process. The right side represents the fabric of isolated scientific facts held in partial form by beginning students. The left half represents the full fabric of science, of which the right is a partial picture. On the right side, levels of facts that are tied together as conclusions rooted in other knowledge are labeled F. Other facts that are learned by students only in an isolated, passive manner are labeled F_A. These are intellectual anomalies, for they lie outside the logical world of related information and least astonishment patterns. Students who accept and memorize such facts are reacting as dogs do to the contradiction of housebreaking. However, students who view these facts as anomalies

46

Application of the
ideas to the problem of
individual learning

and search for additional information and for reasons to tie them together are functioning as scientists. There is a vast difference between a thinker and a knower. The difference involves internal rather than external motivation. Darwin proved it!

The next three chapters of this book are devoted to presenting a special set of tools by which students may form the fabric of geology. Since geology is a fairly representative qualitative science at the level under discussion, the tools and reasoning methods may be useful for students specializing in other disciplines too.

ANNOTATED REFERENCES

1. de Beer, Sir Gavin. 1965. Charles Darwin. Anchor Books, Natural History Library, Doubleday & Co., Garden City, N. Y. 295 pp. (The material quoted begins on p. 30. Darwin's remarks are quoted as they appear in this book, although they originally appeared in different works written by Darwin and used by de Beer in compiling his text. Used by permission of Doubleday & Co.)

2. Davies, Gordon L. 1968. The tour of the British Isles made by Louis Agassiz in 1840. Ann. Sci. **24**:131-146.

3. Hansen, Bert. 1970. The early history of glacial theory in British geology. J. Glaciol. **9**:135-141.

4. Lyell, Sir Charles. 1830. Principles of geology. Vol. I. John Murray, London. 442 pp. (Lyell's awareness of changes in animals through geologic time is evident in the subheadings of Chapter 9, p. 144. A significant number of other available sources indicate that the basic concepts of changes in life forms were well known and appreciated by the scientific community before Darwin published the *Origin of Species.* Darwin's most significant contribution lay in his development of the theory of natural selection, a mechanism by which the changes he described could have taken place.)

PART III

The special characteristics of geology as an historical science

Chapter 4

What seest thou else
In the dark and backward abysm of time?

The Tempest, William Shakespeare

The wasness of the is

The remarks of Jonas Moore to Samuel Pepys have never been included as part of the history of geology. We cannot discuss them with authority but we can enjoy them at face value. Pepys' account contains some astonishing phrases: "England and France were once part of the same continent . . . by very good arguments . . . the time therein is not well computed nor understood."[1] The word "same" implies that Moore had made a comparison of the two areas. We know that Moore began his career in geologic engineering in the seventeenth-century effort to drain the Fens. This low marsh occupies an extensive area in the eastern part of England and is floored with silts and glacial peat bogs. The surrounding rocks on higher ground do have counterparts in France dating from the Jurassic, Cretaceous, and early Tertiary periods (180 to about 36 million years ago). Moore became Surveyor General of Ordnance two years after talking with Pepys. Prior to this he had apparently made some coastal surveys. It is not known whether he had made his own geologic observations and correlations or whether he obtained his ideas from the Dutch engineers who were also employed on the Fens project. Whatever the source, he must have appreciated "some good arguments" and had a remarkable awareness of the meaning of prehistoric time.[1]

Geology is unique among the physical sciences in that it is an historical study as well as a study of the constantly changing earth. The nature of time must be appreciated before any real sense can be made of history and chronology. There are three major **49**

50

*The special character-
istics of geology as an
historical science*

aspects of time that must be separated to avoid ambiguity. The first is time as a dimension used by physicists in the centimeter-gram-second system. Their unit was originally the second, $1/24 \times 1/60 \times 1/60$ or $1/86,400$ of a mean solar day. The mean solar day is the average length of a solar day, which is the average time it takes the earth to revolve so that the sun is directly overhead (at noon) again. Thus the unit of time is an interval between events. The ticktocks, ticktocks of our clocks are artificial divisions that are much easier to use than the spinning of the earth and the appearance of the sun precisely overhead as it passes through a vertical north-south plane. As clocks became more precise, the variations in the spin of the earth became apparent, and the mean solar day became too inaccurate. Ephemeris time was established with the unit called the ephemeris second, defined as $1/31,566$, 925.9747 of the tropical year 1900. Improved clock design has produced a new standard, the atomic second. This is the length of time necessary for $9,192,631,770$ cyclic vibrations of the atoms in a cesium atomic clock. Physicists are also concerned with the flow of time, and their observations from distant stellar sources leave them with uncertain answers to questions about absolute standards. They can measure the second to almost 1 part in 100 million but are not sure that this is an invariable standard. We shall leave them with their perplexing thoughts and Thomas Mann's admonition, "Time has no divisions to mark its passage. . . ."[2] The physicist's view of time is not enough.

The second aspect of time that must be considered to avoid ambiguity is the nature of man's perception of it. Philosophers have worried over this problem for centuries. We shall examine what has been said on the subject of perception as a background to the third major aspect of time, the geologist's view of it as reconstructed from rocks.

Man's perception of time is notoriously relative. Students are aware of this as they wait for a lecturer to stop talking and release them from class. Eric Hoffer may have spent little time in school listening to lectures, but such experiences may have been in his mind when he wrote, "It is waiting that gives weight to time."[3] Once Grope found a cross carved on a desk in his lecture room. Under it was the humbling statement, "In memory of those who died . . . waiting for the bell." William James is among the best-known philosophers who have examined man's perception of time and its relative character. He found that the universal feeling for time is a function of memory, subject to some sort of "law of contrast" associated with the age of the individual: "The same space of time seems shorter as we grow older—that is, the days, the months, and the years do so; whether the hours do so is doubtful, and the minutes and seconds to all appearance remain about the same."[4] Then, quoting from the nineteenth-century work of Professor Paul Janet, James continues.

... the apparent length of an interval at a given epoch of a man's life is proportional to the total length of the life itself. A child of 10 feels a year as 1/10 of his whole life—a man of 50 as 1/50, the whole life meanwhile preserving a constant length. This formula roughly expresses the phenomena, it is true, but cannot possibly be an elementary psychic law; and it is certain that, in great part at least, the foreshortening of the years as we grow older is due to the monotony of memory's content, and the consequent simplification of the backward-glancing view.[4]

James Harris, an engineer, calls this "a logarithmic view of time."

The philosophers have given us a tremendous insight by showing that the perception of time is a function of memory. However, the human life-span is too short to serve as an adequate base for memory. Even the overly mellow Grope, who as a small boy once heard General Nelson A. Miles describe the struggle of stone-age man in conflict with civilization on the western plains, had no contact with Dan'l Boone, much less Cretaceous meadows and grazing dinosaurs. If we are to gain a view of geologic history, the knowledge of events and the intervals between them must be sought another way. The philosophers are of little help at this level, for the records we seek are not written in the language they use. Langdon Smith put it very nicely in his poem "Evolution."

And that was a million years ago,
In a time that no mans knows;
Yet here to-night in the mellow light,
We sit at Delmonico's.

Part of the difficulty of gaining a personal view of time long past is the haunting thought that these wonder stories are just science fiction. Grope once asked a specialist in American colonial history about his experience with time and what made the study a living discipline to him. The man admitted that history had

52

*The special character-
istics of geology as an
historical science*

Fig. 5. Music illustrates the concept of time as the duration between events.

Fig. 6. The wasness of the is. This is a direct print of the grain of a piece of redwood. There are 243 annual rings, indicating that the wood at the lower left is two centuries older than the wood at the upper right. This piece of redwood was taken from the interior of a much larger and therefore much older tree. The width of the rings records variations in climate that affected the rate of the tree's growth. Doesn't the "is" seem less important than the "was"?

little meaning while he was an undergraduate at Sewanee, despite
a Phi Beta Kappa scholastic record. Later his dissertation research
took him to Charleston, South Carolina, for an examination of
original documents. One night after a social engagement he strolled
back to his hotel through the deserted streets in the old part of
town. He walked between the silent walls of handmade brick. The
homes are perfectly kept despite their age and the intensity X
earthquake of 1886. As the trees bent to the sea wind, he saw
black shadows play across white walls as they have done for nearly
three centuries. The smell of drains heightened his sensitivity to
the past, as did glimpses of formal gardens through iron gates.
Even the cobblestone lanes in midblock spoke clearly of former
times, for the quarters in which slaves once lived can be reached
in no other way. Then Grope's friend sharply broke off his thoughts.
"That is how I became an antiquarian. It took discipline to make
me an historian."

The point has always impressed Professor Grope, for geologists
know that the past may be read from the rocks in only one way,
through disciplined experiences. Grope's own oblique approach to
the realization of time requires the expansion of this principle. In
Fig. 5 a drawing is used to convey the meaning of time. The illustration is raw data—lines, dots, and symbols on paper. At the
top, seven sounds are displayed on a musical staff. If played simultaneously as one event, these notes produce the raucous noise
of a single unmusical chord. However, they become the hauntingly
beautiful theme of the folk song "Shenandoah" if they are played
in sequence as shown in the lower diagram. The difference is the
result of changing pandemonium into a disciplined sequence of
time-spaced events. Music is more than sound and more than
rhythm. It is a disciplined interplay of the two. Without a sense
of time and order, music does not exist. The method of depicting
time in Fig. 5 involves the use of symbols to represent events
separated by the quality of duration we call time. This same method
may be taken out of the realm of art and used just as effectively
in science. All that is needed is to place the raw data in the proper
language.

Fig. 6 is a direct print made from the curved grain of a piece
of redwood. An undisciplined view of it is comparable to the top
diagram in Fig. 5. Everything that is to be discussed is there but
hidden in the wholeness of it. Suppose we extract a series of facts
and conclusions based on the simplest sort of observation and the

54

*The special character-
istics of geology as an
historical science*

principle of least astonishment. When we have finished, we shall
have learned how nature retains a memory pattern of events and
the intervals between them. This is not man's memory but nature's,
yet it can become part of our own disciplined experience. Such
are the wonders of the human mind.

First note that the light and dark lines are actually pairs of
annual growth rings. Each pair is a record of the lateral expansion
of the tree during one growing season. If this were a study in
technical botany instead of time, we would have to learn all about
xylem, phloem, cambium, and the various plant fluids. However,
since the two-ringed pattern of yearly annual growth in woody
plants is common knowledge, it is sufficient to recognize the rings
and proceed. The original block from which the print was made
contained 243 pairs of rings. A few were too fine to show up in
the print. The lower left-hand corner of the block is more than
two centuries older than the upper right-hand corner. The curvature
of the rings gives us an additional clue as to their age. The diameter of the tree at this point must have been between two and
three feet. Most commercial redwood comes from large trees, so
we may assume with reasonable certainty that this block was
originally in the interior of the tree and not on the outer rim. If
this is true it is a real antique! We may also assume that the tree
was cut down in the same year the outer rings were formed. If
there was a foot or two of wood between the outer rings and the
piece pictured in Fig. 6, the missing section must represent the last
1,000 to 2,000 years—perhaps more.

Examine the print again and note that some of the groups of
rings are close together and others are farther apart. Although
botanists might wish us to be more precise, it is safe to conclude
that wide spacings represent climatic conditions favorable to growth
and close spacings represent less favorable conditions. Warm, moist
springs and summers are contrasted with times of extended winters
and cool, dry summers. We have just unraveled a part of nature's
memory through a summation of facts and the least astonishment
principle. We know a little more about the climatic conditions
that occurred somewhere in northern California during a span of
243 years some time between the birth of Christ and the Norman
Conquest of England. This is an illustration of one means by which
the memory of man can be extended through disciplined experience. All we need to do is find the data and know how to read
the language.

Perhaps another illustration will serve to show how the method can be extended into the most remote crannies of nature's memory. Fig. 7 is a picture of the Egyptian obelisk that stands behind the Metropolitan Museum of Art in Central Park in New York City. It is 69 feet tall, 7.5 feet wide at the base, and weighs 224 tons. The monument was carved out of a reddish-colored granite near Aswan at the First Cataract of the Nile about 1450 B.C. in celebration of a festival during the reign of the Pharaoh Thutmose III. The Egyptians floated it many hundreds of miles down the Nile and erected it at Heliopolis, a city at the head of the delta where the river divides into a number of distributary streams. Some fourteen centuries later, in 22 B.C., it was moved again, to Alexandria at the mouth of the westernmost distributary of the Nile. It was erected before the Caesareum at Nikopolis, a suburb of the main city. In 1880 the Egyptian government presented it to the people of the United States. The great obelisk, supported by four bronze frogs at the corners, has looked down on a truly foreign world since 1881.

There is more to the story than this. We can read it much as we can read the record of a life in the lines of an old man's face. The tale begins with the origin of the rock in late Ordovician or Silurian times, no less than 410 million years ago. Not all granites are exactly alike, but in general they have been formed by the partial melting and recrystallization of older rocks. The process probably occurs at depths of 8 to 12 miles below the earth's surface, where slow cooling permits the formation of large crystals. The coarsely crystalline structure of the Aswan granite is very obvious when the monument is examined closely. Time is known only as the duration between events. We have added another tick, tock to the story so we know more about the time it represents. The tick occurred 410 million years ago with the origin of the granite, the tock when the Egyptians carved the stone. In the interval between tick and tock, erosion by rivers more ancient than the Nile were able to strip away the 8 to 12 miles of overburden and expose the granite for quarrying by the ancient Egyptians. There are many other events that took place in the Aswan area during the same time span, and a recorded memory of them is part of the field evidence in Egypt and the Sudan, but the tourist in Central Park cannot see them at this distance. All he has before him is the silent obelisk; yet, if he knows the language even this small sample has a great deal more to tell.

56

*The special character-
istics of geology as an
historical science*

The hieroglyphs are sharper on one side of the stone than the other. Fig. 7 shows quite clearly that the carvings have been beveled from one edge and part of one side. Wind erosion is responsible for this. At some time in the past, either at Heliopolis or at Alexandria, the monument must have fallen over in the soft sand without breaking. The desert winds, driving sand particles as cutting tools, were able to polish away some of the unprotected surfaces. Such things do not happen in Central Park. Weathering there is of a chemical nature. The cold, wet climate of New York has taken its toll, for frost heaving has begun to loosen particles and damage the surface. All the minerals in granite were formed in chemical equilibrium at high temperatures and pressures far beneath the surface of the earth. Most of them are not in equilibrium with atmospheric reagents, free oxygen, carbon dioxide, water, and the sulfur dioxide of big-city smog. Chemical and mechanical weathering have damaged the obelisk a great deal more during the short time that it has been in New York than these same agents ever damaged it during the millenia spent in Egypt's dry climate.

Fig. 7. The Egyptian obelisk in Central Park, New York City. Another study of the wasness of the is.

This particular illustration has been chosen because the conclusions range through time and changes of climate that are all deduced from factual observations. The history of this one granite obelisk encompasses half the earth, hundreds of millions of years before man's entrance on the stage, the full span of civilization, and ends in the backyard of the Metropolitan Museum of Art. It can be fun to visit the Metropolitan, for the same sorts of analyses may be made of a number of its displays. Just as an example, the iron in the medieval armor comes from the interiors of forgotten stars.

The average tourist misses these things because he has not been disciplined to think in terms of well-ordered time. Most men know only two categories of time, *"is"* and *"was."* The "is" category is clear enough, but the "was" category is a jumble. The old buildings of Europe are pointed out in a manner that separates them from the social and historical context that gives them meaning, although a date is attached almost as if it were a price tag. Prehistory is altogether out of range of most tourists' imaginations.

The concept of the "wasness of the is" has been used but unnamed in both these illustrations. The obelisk "is" visible in Central Park. The piece of redwood from which the print has been made "is" a reality. In both cases details from the ordered past may be read directly from the data in the objects themselves. Their wasness remains with them, to be reread as disciplined experiences. This is something of what Alexander Pope (1688-1744) meant when he wrote: "For he lives twice who can at once employ/The present well, and ev'n the past enjoy."[5]

This technique of ordering events is a tool with which it is possible to understand the realities of past time. Without it the view of the world is unnaturally constrained to the absolute now, a present without a history. Once the idea of thinking in terms of the "wasness of the is" becomes a habit, every rock and landform take on new meanings. They become pages in the books of earth history.

There are additional techniques that must also be mastered before the events that divide past time become apparent. One technique is called the "cone of vision concept." It helps the geologist to place many aspects of nature simultaneously on his mental viewing screen. This technique is discussed in Chapter 5.

Still another method of seeing the world is explained in Chapter 6. This is probably the most important chapter in the book, for

58

*The special character-
istics of geology as an
historical science*

it is an explanation of the way in which the what, when, where, which, how, and why of earth science all fit together.

When Professor Grope teaches this material, he cautions his students to be patient. At first the ideas may seem overly complicated, but they will be straightened out clearly enough. One freshman was heard to mutter, "You can say the same thing about a tangled fishing line, but it is more fun to fish than to untangle knots!"

ANNOTATED REFERENCES

1. Stephen, Leslie, and Lee, Sidney. 1967. The dictionary of biography. Vol. XIII. Oxford University Press, London, 1359 pp. (The information on Jonas Moore is found on pp. 820-821.)
2. Mann, Thomas. 1938. The magic mountain. Alfred A. Knopf, Inc., New York. 433 pp. (This familiar line, which appears on p. 287, is part of a description of the artificial boundaries that men place on the flow of time.)
3. Hoffer, Eric. 1955. The passionate state of mind. Harper & Row, Publishers, New York. 151 pp. (This quotation is taken from p. 146. Used by permission of Harper & Row, Publishers.)
4. James, William. 1950. Principles of psychology. Dover Publications, Inc., New York. 869 pp. (James' view of time is taken from Chapter 15, p. 605 ff. Used by permission of Dover Publications, Inc.)
5. Pope, Alexander. Imitation of Martial. In Ault, Norman (ed.). 1954. Minor poems. Methuen & Co., Ltd., London. 492 pp. (These lines, Book 10, Epigram 23, are found on p. 167.)

Chapter 5

It lies around us like a cloud,
A world we do not see . . .

The Other World, Harriet Beecher Stowe

The cone of vision concept

The human eye is located at the apex of a cone made of light moving toward it. The physical limits of the large end, or base of the cone, are the various sources of light. When we look toward the stars in a clear night sky, this limit is so far away that it is measured in light-years, and the distance across the open end of the cone is unbelievably vast. If we look at the ground on which we walk, the length of the cone is only a few feet, and the width of it is correspondingly reduced. Each of us would see exactly the same things if the human eye were no more than a camera. What we do see is controlled by a filter system that the mind uses to screen out some impressions and heighten others. *The cone of vision concept is a means of heightening impressions that pertain to geology or to some other special field.*

We wish to be able to make local observations in as wide a range of categories as possible and see them in a broad context rather than as isolated, close-up views. Fig. 8 shows how the cone of vision concept is used. In this figure a geology student is holding up a hand specimen or rock so that its texture parallels that of the cliff in the distance. The minerals in the small piece of rock are minute rods and plates and the piece has a banded appearance. The massive rock in the distant cliff is made of the same minerals in the same forms, so that the cliff is also banded. The student cannot see the minute bands at this distance, but he has climbed on the cliff and knows they are there. He can see large structures that have been formed as weathering has etched out the resistant and nonresistant layers produced by minor variations in the com-   59

60

*The special character-
istics of geology as an
historical science*

position of the rock. The student is also aware that the rocks of the entire region for miles around are similar to those shown in Fig. 8. This means that he has already made some additional observations that permit him to conclude that both the hand specimen and cliff rocks are representative samples of the region.

The value of the cone of vision concept is beginning to be apparent. From this one spot the view for the student is no greater than it is for the average tourist. However, there is a filter system in the student's mind that allows him to see the cliff as a part of the larger structure. The tourist cannot do this. With an untrained mind, all the tourist can see is a mountain. Personal experience produces a filter system through which the world is viewed. Specialized education builds the technical filters based on technical experiences. It is the same perspective we gained in Chapter 4 with regard to experiencing time.

The student has been taught several other things about the rocks of the area, and these, too, help him to see more perceptively.

Originally
the same
mountains

Fig. 8. The cone of vision concept.

The texture that is so obvious in the hand specimen was formed

with the rock about 360 million years ago. Measurements of the
alignment of the plates show that they extend in a northeast to
southwest direction. This is exactly the same orientation of the
Appalachian Mountain system, of which the rock is a part. Interestingly enough, there is a similar set of rocks and mountains
of the same age crossing parts of Ireland, England, Scotland, and
the length of Norway. Before continental displacement occurred
about 180 million years ago, all of these mountains in Europe and
North America were parts of a single continuous chain. They fit
together as well as parts of a torn newspaper that may be read if
the edges are rejoined. Therefore the student sees beyond the
cliff along the line of the Appalachians over to Norway. His technical experience permits him to focus the whole story on his cone
of vision screen, unimpeded by continental displacement. He is
not confused by the presence of 3,000 miles of ocean because his
cone of vision perception involves a sense of time. The Appalachians were made in stages; principal episodes occurred about 360
and 250 million years ago. The younger one trends off through
Central rather than Northern Europe. The student has learned to
see all of this as it once stood for fully 70 million years before the
continents were pulled apart.

Vision of things beyond the horizon must be displayed on some
sort of mental screen, just as a movie must be displayed on a screen
in order to be made visible. In this case we will use the map of the
world as our screen. The northeast-to-southwest orientation of both
the Appalachian Mountains and the backbone of the peninsula of
Norway and Sweden does not have to be emphasized. Geologists
use standard base maps and display all sorts of specialized data on
them. The maps can then be published and used as screens for
the communication of ideas to others.

A history of map making is in a very real way part of the history or the cone of vision concept. Maps have been used for several
thousands of years as a means of communicating patterns of reality
beyond the immediate limits of natural sight. We shall gain our
own perspective of this history by comparing the work of men
of different ages, peering back and forth in time. This is not good
historical methodology, but it fulfills our purpose just as counter-
point serves in a musical composition. We wish to see how geographic mapping serves as a screen that will allow us to hold the
world up to view. We can develop this theme best by setting minor

62

*The special character-
istics of geology as an
historical science*

themes against it, highlighting the realities with contrasts and similarities.

The most primitive mapping is probably done by animals that stake out the metes and bounds of their territories with roars, howls, songs, and scented sprays. Psychologists call this "species-specific behavior" and accept it as a universal part of nature. Groups of animals are food-gathering societies. Each unit needs enough territory to ensure an adequate food supply. The howls and yowls and songs are true maps communicated to other members of the same species; they are either accepted by tacit agreement or are the basis for open conflict. The most primitive human societies that depend on food gathering function in a similar manner.

The point at which mapping became an abstraction in picture form is lost in antiquity. Voute[1] cites a geologic mining map drawn on a papyrus scroll that dates from the period of the Egyptian Pharaoh Seti (1313-1292 B.C.). The date places it in the Bronze Age, so this map is not likely to be the first abstraction, although it may be the oldest surviving map. Complex irrigation systems date from much earlier times. These must have been mapped as part of the planning procedure and also as a means of distinguishing valuable ownerships. Simple maps drawn with sticks on bare ground as far back as 6000 B.C. are not beyond speculation. A day's work could be laid out in this way so that each worker would know what to do. Similar maps for hunts and military raids may date as far back as the time of Neanderthal man. We have no way of knowing.

Contemporary maps share three distinguishing characteristics. They are drawn to scale, so that distances between the various points on the map are proportional to the actual distances between the corresponding points on the ground. They are oriented with respect to north, and all the directions on them correspond in some way to directions that have been determined by ground surveys. Finally, they are made to represent portions of a sphere. This is accomplished in a number of ways. The simplest is to make a spherical map, or globe. Portions of the globe are reproduced on flat pieces of paper in a variety of projections, each designed to preserve some aspect of orientation and proportion.

The history of ancient map making must be seen in terms of the development of technology and means of making accurate surveys, as well as in relation to the tremendous accomplishment involved in the discovery that the earth is a sphere. Travel beyond the immediate horizon of home territory opened the way to an expanded cone of vision. We still think in these terms when we remark that "travel is broadening." Ancient overland travel presented a slowly unrolling landscape, with the resultant illusion that the earth was a flat surface. Mapping was based on the principle of adjacency. Points were placed in reference to other points in a continuous chain, just as they were seen by travelers.

A classic example of this type of mapping is the famous Roman road map of the first century A.D.[2] This map is on a scroll 21 feet long and 1 foot wide. Thousands of miles of paved Roman roads are shown on this one strip. The center section represents the length of Italy and all the roads are strung out without regard to direction. Distances are marked between intersections. Spaces representing the Adriatic Sea and the Mediterranean Sea are left on either side of the Italian strip. Then the roads through southeast Europe and on to the area of the Holy Land are shown stretched out as are those of Italy. Beyond the Mediterranean the roads of North Africa are also drawn so that they connect in the area of Egypt with those of the Holy Land.

It is a peculiar-looking map, for, aside from the attempt to set out a view of position by adjacency and some representation of distance along the roads, it is not similar to our maps. The map had one important value—it worked. The scroll was easy to carry about. A spread-out map with this amount of detail would be very difficult to handle in the wind. A traveler could easily tell where he was and what roads he should take to make a given journey. One look at the map and the familiar saying "All roads lead to Rome" comes to mind immediately, for that is exactly how this map is drawn. Rome is at the center of everything.

World maps were already in existence by the time of Herodotus (485-425 B.C.). Although they were drawn with a good deal more imagination than fact and set out solely on the principle of adjacency, they do represent an expanded cone of vision. In one way they were less stylized than the Roman scroll maps, because there was an attempt to show the features in some pattern of orientation. Herodotus had little use for them because he knew that the play of imagination instead of fact was a poor base for improvement. His statement is worth repeating.

For my part, I cannot but laugh when I see numbers of persons drawing maps of the world without having any reason to guide them;

64

The special character-
istics of geology as an
historical science

making, as they do, the ocean-stream to run all around the earth, and the earth itself to be an exact circle, as if described by a pair of compasses, with Europe and Asia just of the same size.[3]

Herodotus was a great traveler and he understood the principle of adjacency. Apparently the maps that showed a limited earth bothered him most, for he had talked with people from distant lands and knew that no one had seen such a limit. His own view of the world was unbounded, with an undefined Europe to the north and an undefined India and Asia to the east. He reported that Africa had been circumnavigated by some Phoenician sailors during the reign of the Pharaoh Necos some 200 years earlier (circa 600 b.c.). Their ships had been dispatched from the Gulf of Suez with orders to sail south. They returned a few years later through the Strait of Gibraltar and the Mediterranean Sea to Egypt. The trip might be suspect were it not for a singular fact that Herodotus reported in disbelief:

> On their return they declared—I for my part do not believe them, but perhaps others may—that in sailing around Lybia [Africa] they had the sun [at noon] upon their right hand. In this way was the extent of Lybia first discovered.[3]

The timeless value of a good observation is demonstrated in the report. Herodotus did not know, of course, that the earth was a sphere orbiting the sun and that the Phoenicians had twice crossed the equator and the plane of the ecliptic. The position of Africa is such that the Phoenicians saw the sun at noon on their right hand while sailing west around the Cape of Good Hope. At that time they were south of the equator. After turning east through the Strait of Gibraltar, they were north of the equator and saw the sun once more on their right at noon.

Information that to us is within the normal range of fact and a situation of least astonishment was to Herodotus beyond the range of credibility. Some historians downgrade Herodotus for his tendency to report any information given to him without making personal judgments. Geologists feel that this is an admirable quality, because it gives future generations a better data field with which to work. The voyage of the Phoenicians was a tremendous achievement that was not equaled until 2,000 years later, when Vasco

da Gama reached India by circumnavigating the Cape of Good
Hope in 1498.

Ancient navigation, like all other travel in those days, was first based on the principle of adjacency. The Phoenicians practiced it by sailing along the coast of Africa. Their success was due to the insular nature of the African continent. How different the result would have been had they attempted to sail around Asia and reached the Malay Peninsula and the island arc of Sumatra and Java. An example of this "confused" type of navigation may be drawn from Homer's *Odyssey*. Coasting along a familiar shore and recognizing points and inlets as one does old friends sets restricted limits on the range of exploration. Whether the voyage of Ulysses is purely fictional or whether Homer drew on accounts of real sailors is not too important. Our interest is in the development of the cone of vision concept and how the terms "lost" and "not lost" relate to it.

In Homer's time (circa 750 B.C.) the Mediterranean Sea was not completely charted. Ulysses is said to have left Troy about the beginning of the twelfth century B.C. with the intention of sailing around the northern rim of the Aegean Sea and coasting down the shore of Greece to the Peloponnesos and Penelope. He could have sailed directly west into the setting sun and then gone down the coast to the south if he had chosen to venture into this open water. According to Bradford, Ulysses sailed first to the north and west to pick up a landfall and make a raid on the Cinonian coastline. Skipping the events of storm and island contact on the next leg of the voyage, we find Ulysses and his crew caught in a severe *Grego*, the Greek wind, and driven southwest beyond Cythera, the last of the Greek islands. "For nine days I was driven by those cursed winds across the fish-infested sea."[4] Then on the tenth day they reached the Land of the Lotus Eaters, which Bradford places on the Island of Jerba off the coast of Tunisia in North Africa. From here Ulysses does not sail far enough to the northeast, and he is then in Homeric trouble!

The importance of the cone of vision concept is evident in a comparison of the voyages of Ulysses with those of Lindbergh and Thor Heyerdahl. Both modern Argonauts reached distant goals by knowing where to find them. Lindbergh was lost for a time in a fog off the coast of Ireland and was fortunate to be directed to land by some fishermen who pointed the way. At sea the principle of adjacency is of no value; it is necessary to navigate in terms of

66

*The special character-
istics of geology as an
historical science*

direction and distance. Lindbergh used his compass for direction and estimated distance on the bases of airspeed and time. Heyerdahl had no control of direction because of the nature of his experiments, but he was able to measure distance in terms of changes in longitude and latitude on an earth of known shape and size. It took a long time for man to learn to measure the three parameters—direction, latitude, and longitude—at sea. The beginning of the story is lost in time, and the end is not fixed until 1775. Placing the earth in the cone of vision was a hard-won victory!

The use of the sun and stars to indicate direction is a technique whose history is far too ancient to trace. Devices for measuring angles are also very old. Some of the precision that was obtained without the use of lenses or verniers is simply amazing. The miracles of orientation found at Cheops' Great Pyramid in Egypt (circa 2700 B.C.) and at Stonehenge in England (circa 1800 B.C.) attest to the skill of their architects and engineers. We may assume that the early navigators could set a course and hold it with the help of the sun and stars in good weather and the ever-welcome landfall.

Pytheas, in about 225 B.C., was the first to conceptualize and then prove that the earth is a sphere. He traveled to different places and measured angles in a vertical plane between a plumb line that served as a reference and sightings on the same stars. From his data he concluded that the earth had to be a sphere. His conclusion is obvious, because the angle to a star should vary at different latitudes. It is also obvious that sailors familiar with the manner in which ships appeared and disappeared over the horizon could have recognized the spherical shape of the earth centuries before Pytheas. Perhaps they did and have simply received no credit.

Eratosthenes, the librarian at Alexandria, used a similar system and actually determined the size of the spherical earth at about the same time. He knew from surveys that had been made by Egyptian engineers that the horizontal distance from a point on the upper Nile near modern Aswan to Alexandria was about 500 miles (our units) along a very nearly north-south great circle line. The angles that he measured indicated that this distance was 1/50 of the circumference of the earth. Thus the total circumference measured 25,000 miles. An erroneous measurement made by Posidonius in the first century B.C. reduced this figure to 18,000 miles. Strabo (63? B.C. to after 21 A.D.) used Posidonius' figure in his *Geographia*. The same error was included on Ptolemy's maps of the second century A.D., and the map makers of the Renaissance reprinted it. Finally, 1,500 years later, Posidonius' mistake was responsible for Columbus' belief that he had found the Orient instead of the New World. Columbus was under the impression that the earth was a much smaller ball than it is.

Another Greek, Hipparchus, proposed a network of latitude and longitude lines forming a 360-degree grid to cover the whole earth in about 130 B.C. Such a scheme was not completed and adopted on a worldwide basis until 1884. There were many reasons for the delay. Although latitude could be measured fairly easily, even from a pitching ship at sea, through the use of the progressively more sophisticated kamal, astrolabe, quadrant, backstaff, and finally the sextant, measurements of longitude were very difficult to make. One degree of longitude is the amount of the earth's rotation in four minutes of time. In order to be able to measure longitude, the time differences between some chosen base point such as Greenwich, England, and a second point must be known. Local time can always be determined by "shooting the sun" at noon, but how could the early explorers know what time it was "back at the ranch"? Finally, in 1714 England's Board of Longitude offered a prize of 20,000 pounds to anyone who could construct a mechanical clock so accurate that it could be set according to Greenwich time and carried on shipboard for use on long voyages without appreciable error. The prize was belatedly given to John Harrison in 1775 for a chronometer he had invented in 1736. Even on shipboard his mechanism was accurate to one second a month! Captain James Cook used Harrison's chronometer (model No. 4) on his voyage around the world in 1772 with great success. Even then the intervention of King George III was required before the Board of Longitude paid Harrison his prize money, which he received shortly before his death. A century after it became possible to measure longitude accurately, national pride still postponed the adoption of a worldwide system. Each major country wanted the base longitude to pass through its own capital.

The magnetic compass became available for determining direction in the twelfth century. Ship design had been improved by that time, and exploring navigators were capable of setting a course and sailing over great distances with more safety than they had ever known. The measurement of distance at sea had always been an uncertain affair. A floating anchor, attached to the ship by a line, could be dropped overboard, and the rate at which the ship

68

The special character-
istics of geology an an
historical science

moved away from the anchor yielded a fair estimate of sailing speed for a short interval of time. Estimation of sailing speed was made uncertain by the effect of currents, so that measurement depended as much on judgment as on engineering.

After 1780, when the chronometer became generally available, daily fixes of latitude and longitude permitted navigators to plot their ships' courses with a precision that had never been known. At last the sailors could know where the ship was with respect to the cone of vision picture of the world. The only major uncertainties left were created by bad weather when the navigators could not see the sun and by the effects of uncharted currents. Matthew Fontaine Maury pioneered in the work of charting the ocean currents in the mid-nineteenth century. The following quotation from the introduction to his 1855 *Treatise on the Physical Geography of the Sea* is a clear statement of the use of the cone of vision concept.

The primary object of "The Wind and Current Charts," out of which has grown this Treatise on the Physical Geography of the Sea, was to collect the experience of every navigator as to the winds and currents of the ocean, to discuss his observations upon them, and then to present the world with the results on charts for the improvement of commerce and navigation.

By putting down on a chart the tracks of many vessels on the same voyage, but at different times, in different years, and during all seasons, and by projecting along each track the winds and currents daily encountered, it was plain that navigators hereafter, by consulting this chart, would have for their guide the results of the combined experience of all those whose tracks were thus pointed out.

Perhaps it might be the first voyage of a young navigator to the given port, when his own personal experience of the winds to be expected, the currents to be encountered by the way, would itself be blank. If so, there would be the wind and current chart. It would spread out before him the tracks of a thousand vessels that had preceded him on the same voyage, wherever it might be, and that too, at the same season of the year. Such a chart, it was held, would show him not only the tracks of the vessels, but the experience also of each master as to the winds and currents by the way, the temperature of the ocean, and the variation of the needle. All of this could be taken in at a glance, and thus the young mariner, instead of groping his way along until the lights of experience should come to him by the slow teachings of the dearest of all schools, would here find, at once, that he had already the experience of a thousand navigators to guide him on his voyage. He might, therefore, set out upon his first voyage with as much confidence in his knowledge as to the winds and currents he might expect to meet with, as though he himself had already been that way a thousand times before.[5]

The air age brought new navigational problems, for airspeeds were too great for the standard techniques to be of help. The principle of adjacency and dead reckoning were all that were available for the first three decades of air travel. Pilots followed railroads and rivers or made sightings of towns, mountain peaks, and shoreline characteristics, just as the sailors of Ulysses' time had navigated the Mediterranean. Airspeed and compass direction were the two tools for dead reckoning. Pilots were able to estimate their wind drift from the position of the axis of the plane with respect to the flight path across the ground. Finally, in the 1930s, radio navigation became a reality. The cone of vision was attuned to supersonic speeds. It is never impossible to reach a destination if you know where you are, have the equipment, and know what the environment is like.

It is interesting to contrast the cone of vision thinking that is part of our technical society with the haphazard and unplanned pattern of the preceding millenia. Today we want to expand our cone of vision to develop an adequate picture of the near planets and of the mechanisms of creation. Therefore we set up agencies to do it. They are properly funded, and after developmental procedures are completed, we put some skilled teams of pilot-observers in suitable space vehicles and send them out to find answers.

How different this is from the path of history prior to the second half of the twentieth century! Consider these dates and events, noting the time lapses between them. In 1492 Columbus brought a personal dream to reality. By 1542 the Spanish had penetrated into the western United States as far as the Grand Canyon in Arizona. In 1607, exactly 115 years after Columbus, the English established their own first permanent settlement in the New World. In 1803, almost two centuries later and more than a quarter of a century after the voyage of Captain Cook with the Harrison chronometer, President Thomas Jefferson felt it was time to explore the newly purchased western lands. Two hundred and sixty-three years after Coronado searched Arizona for the Seven Cities of Cibola, we finally sent Lewis and Clark out west to see what was there. Has man become more adventurous, or is his new curiosity a matter of available wealth and horsepower?

70

The special character-
istics of geology as an
historical science

We have seen something of the struggle to put together a view of the world. Now it will be interesting to test the cone of vision concept as a means of gaining insight. Two simple historical situations will serve to relate the saga of man and the earth. First the story of Columbus as it is told in history classes. In the late summer of 1492 his ships were headed for the New World with a rag-tag crew ready to mutiny because they feared falling off the edge of the earth. Columbus said, "Sail on, and on, and on!" They sailed on. The muttering subsided when signs of land were seen, and before the crew lost hope a landfall was made at San Salvador in the Bahamas.

This view of history places too much emphasis on the picture of men in small ships and omits the much larger picture of the moving ocean and atmosphere that contained both ships and crews. What part did the planetary winds and ocean currents play in the story? Columbus and the crew were in ships, but the ships were in the water and atmosphere, at certain latitudes and at certain times.

Sunlight strikes the earth more directly in the equatorial zone than at the poles, making the air hotter and therefore less dense in the tropics. This density variation triggers a flow of denser and

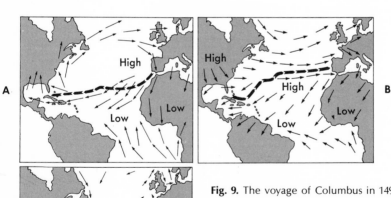

Fig. 9. The voyage of Columbus in 1492 was controlled to a great extent by prevailing winds and ocean currents. **A,** Summer winds. **B,** Winter winds. **C,** Ocean currents.

colder air from the poles toward the equator. The warm air aloft over the equator flows poleward, completing the cycle. The flow is complicated by a number of factors. Two of them are the precession from winter to summer and the rotation of the earth. The rotation of the earth causes the air to drift laterally as well as toward the low-pressure zones near the ground. The trade winds are the result of this drift.

Columbus left Palos, Spain, on August 3, 1942, and sailed to the Canary Islands (Fig. 9). The summer trade winds behind him drove his ships on. Then he turned due west and sailed, with the wind still behind him but slightly to the right (sorry, wind astern and well to starboard), until he was in midocean. His course from there to the landfall was directly downwind, west-southwest. The landing was made on October 12. It is within the realm of logical speculation that his sailors were more afraid of their inability to tack back to Spain against this strong tail wind in their clumsy ships than they were of falling off the edge of the earth. They must have known how much food and water were on board and how long it would take them to get back. Yet Columbus decided to sail—on, and on, and on!

The return voyage was begun from Hispaniola on January 4, 1493. The course lay northeast, following the trade winds and currents for that time of year until the latitude of Spain was reached approximately in midocean. From here the course was due east, with the winter trade winds well behind them. Landfall was made on March 15. The broken lines in Fig. 9 show that Columbus' course was the obvious one to take maximum advantage of the winds.

Another illustration of the value of cone of vision insight may be taken from the popular myths about the agricultural South in the decades before the American Civil War. The South is often viewed as a backward agricultural land unaffected by the Industrial Revolution. The North, on the other hand, is seen as an alert industrial community, somewhat contemptuous of the backward South. What are the geologic facts that in a way contain the two different regions?

Virginia, North Carolina, South Carolina, and Georgia are composed of the same three physiographic provinces. Each has a low, flat coastal plain with rivers flowing on gentle slopes to the sea. The average elevation 100 miles inland is about 300 to 350 feet above sea level. The Piedmont Province in each state is a flat, tilted plain

72

The special character-
istics of geology as an
historical science

that is cut across crystalline rocks and beveled by streams. In South Carolina the elevation at the upper edge of the Piedmont is about 1,000 feet at a distance of 240 miles from the sea. Beyond the Piedmont is the mass of the Appalachian Mountains. The eastern mountain unit, called the Blue Ridge, is also made of crystalline rocks. The western part, found only in Virginia, contains folded sediments with some coals. All the other rocks are completely devoid of coal and major mineral resources. Because the Atlantic Ocean was formed as a result of continental displacement of the New World from the Old World only about 180 million years ago, the Atlantic drainage is younger than the westward drainage into the Gulf of Mexico. Most of the water that flows from the humid Smoky Mountain area in western North Carolina, Georgia, and eastern Tennessee belongs to the old drainage. These were the streams that were tamed by the Tennessee Valley Authority in the 1930s. The eastward-flowing streams have less water and flow down much lower slopes. This is the stage on which the play of Southern history was performed. What does it mean?

The Industrial Revolution occurred as man learned to produce goods with power-driven machines. There were two important sources of power in the early 1800s—steam and water. Steam was generated chiefly through the burning of coal. The South had only one small supply in the western tip of Virginia that could be called commercial. Railroads would eventually bring West Virginia coal to the deep South directly over the mountain barrier. Shipment could not be made by water, for the rivers ran the wrong way. Shipment of coal from West Virginia and Pennsylvania due east to the northern Piedmont and then southwest to Georgia was possible but expensive. Necessities that would have permitted the South to compete with the North were not readily available prior to World War I.

Waterpower is produced as masses of water fall, transferring energy to machines. Two factors control the amount of power available: the volume of water that falls and the height of the fall. New England is a humid region, with streams that tumble out of the northeastern part of the Appalachians on short, steep courses to the sea. There is no broad, low, flat coastal plain to diminish the energy. These are the same mountains that were originally connected to those in Europe (Fig. 8). The New England coast is the place where the mountains were broken apart, so that naturally they are close to the sea. The situation in the South is quite different. The mountains are far back from the sea, and the settled land of the early 1800s lay in the Coastal Plain and Piedmont. There were some places where reasonably high waterfalls were available for power generation. Small cotton mills were set up on them but not to the extent that characterized the North.

The Old South, thwarted in its attempt to develop industry, maintained a traditional agricultural economy. The New South is powered by coal brought directly through the mountains on railroads like the Virginian, the Norfolk and Western, and the Clinchfield. High dams constructed with government aid have made it possible to generate power on the Piedmont streams. The Piedmont, with better power and better transportation, enough people, and less big-city unrest, is becoming a new heartland. To this degree geographic determinism is a factor in modern society.

The cone of vision concept is such a standard part of scientific thinking that it is impossible to write without using it routinely. We have already used it in describing the fabric of science. All the illustrations and local geology in this book are part of it. The concept is so natural that it is rarely considered separately from the details it brings into focus. Our principal reason for stressing it here is to be sure that, as we skip from one part of the world to another, we do not think of the isolation of place as "real." There is an adjacency involved that connects each place to areas that surround it. The old farmer who said, "I do not want all the land in the world, I just want that which adjoins mine," had a proper concept of the cone of vision.

Professor Grope uses two favorite anecdotes to illustrate to his students exactly what the cone of vision concept means. As an undergraduate Grope lived in Richmond, Virginia. One Sunday afternoon he had gone to the Black Heath area in Chesterfield County on the south side of the James River to look for plant fossils in the shales of the Richmond Triassic Basin. The rocks are about 200 million years old and contain a few low-grade coal stringers of a submarginal commercial quality. Optimistic entrepreneurs have attempted to mine the area since colonial times. After poking his head into three or four old colonial mine tunnels that had been exposed by more recent dragline diggers, Grope found a large block of shale that he could split readily with his geologist's pick. He worked very carefully and broke the block in half along a bedding plane, revealing a magnificent bladelike leaf frond some 15 inches long and 1½ inches wide. Fossils are

74

*The special character-
istics of geology as an
historical science*

usually casts or molds made of stony material. This one was different. The actual leaf lay before him as a thin and free film of carbon. For a moment the leaf was perfect. Then the breeze and first exposure to the air in 200 million years began to break it up. What to do? For a moment Grope was paralyzed. Then he had one of the few strokes of pure genius that have ever come to him. Here was a Triassic salad going to waste: So he ate it!

The cone of vision concept should be used as an intellectual filter system to pinpoint important ideas and display them in relation to each other. The concept is involved in the definition of understanding: knowing enough about something so that every part of it is seen in proper relationship to all the other parts and to the whole.

His other anecdote is about a remark made by a freshman during a field trip. On this particular golden afternoon in Indian summer, the class was studying flood plain deposition. Old Grope was leading a long line of mud-plastered students through a swampy area, where ponded water was trapped between the natural levee and valley wall. Somewhere behind him a voice cried in the wilderness. "I feel like a salamander."

Ah-hah! Here was a philosopher, a man with a cone of vision broad enough to include the story of evolution. This man saw himself, as a mammal, linked to his aquatic vertebrate ancestors of the Predevonian period (prior to 400 million years ago). He saw himself in the line—fish, amphibians (salamander), reptiles, mammals—man and student of all that has gone before. Mark him well!

ANNOTATED REFERENCES

1. Voute, Caesar. 1968. The changing art of geological surveying. Uitgeverij Waltman, Delft, The Netherlands. 29 pp. (This famous mining map is mentioned on p. 5. The map itself is in a museum in Turin.)

2. Tooley, R. V., and Bricker, Charles. 1969. A history of cartography. Thames & Hudson, Ltd., London. 276 pp. (A portion of this wonderful Roman road map is reproduced on p. 22.)

3. Komroff, Manuel (ed.). 1928. The history of Herodotus (translated by George Rawlinson). Tudor Publishing Co., New York. 544 pp. (Herodotus' view of maps and of the circumnavigation of Africa were taken from p. 215 ff. Used by permission of Tudor Publishing Co.)

4. Bradford, Ernie. 1963. Ulysses found. Harcourt Brace & World, Inc., New York. 238 pp.

5. Leighly, John (ed.). 1963. The physical geography of the sea and its meterology (by Matthew Fontaine Maury). The Belknap Press of Harvard University Press, Cambridge, Mass. 432 pp. (This quotation appears on p. 3. Used by permission of the Belknap Press of Harvard University Press.)

Chapter 6

One for all and all for one,
That is our device.

The Three Musketeers, Alexander Dumas

The four critical geologic aspects

The cone of vision concept serves as a screen onto which ideas filtered by the human mind may be displayed and viewed. In this chapter we shall develop this filter system and show how it is used. Sherlock Holmes demonstrated that what he saw was identical to what everyone else saw. The difference lay not in his eyesight but in his ability to view data in significant combinations. Holmes was such a master observer that he thought in terms of conclusions rather than facts. We all read written language this way. Instead of seeing letters we use entire words as the smallest unit, but we do not think in terms of words. Our thoughts are couched in idea units, phrases, clauses, sentences, and paragraphs. Good writing may be distinguished from poor writing by the ease with which the thoughts flow, one from the other. There is a method for writing and another for reading. Reading is easier. Once an even rudimentary reading skill is developed, the method is rarely thought about again.

There is a literature of geology spelled out in minerals, fossils, rocks, formations, structures, landforms, and processes. This literature lies around us all the time. We see it everywhere we go. In order to read it, we need only learn something of the methodology. Language skills are no different than other skills. Mastery is an open-ended effort. One can never learn all about anything, but

76

The special character-
istics of geology as an
historical science

even at the beginning level, a little competence will yield very satisfying results. We already know how to assemble isolated facts into conclusions based on the principle of least astonishment. This much of the methodology is behind us. Now we must learn to recognize the significant combinations of data. This is not as hard as it sounds. There are only four combinations, and they grow out of the common questions that are the basis of all good reporting: where, what, which, when, how, and why. In geology the question of who is unimportant. The four aspects are (1) position in space, (2) identification and classification of materials, (3) chronologic relationships, and (4) genetic relationships. Here are the definitions.

1. *Position in space*

Position in space refers to two kinds of spatial relationships. The first places a geologic feature within the major framework of latitude, longitude, and elevation, or within a continent or political division such as a country or state. The second involves its adjacency to other features that can also be identified. We used this concept in identifying the cliff in Fig. 8 as part of the Appalachian Mountain range. Sedimentary rocks are deposited in superposed layers, just as snowfalls are laid down. An observation that places a particular layer at the top, middle, or bottom of the stack is a conclusion regarding the layer's relative age. This is the way Sherlock Holmes thought. He would see the layers of a snowfall in terms of age rather than in terms of position alone. Geologists use the many aspects of relative position so naturally that they rarely think of them as conclusions. They do not consciously observe position alone; they also observe age.

2. *Identification and classification of materials*

The identification and classification of materials are based on a tremendously wide range of specific facts. Materials are described in the standard units that are used by all the other sciences. Minerals are natural chemical compounds. They are studied and described in chemical terms. Rocks are combinations of minerals. Some are also studied in chemical terms and others in physical terms, using all the techniques of those sciences. Fossils are records of former life forms that have been preserved in the rocks. Systems of classification and description drawn from the master sciences are used wherever possible. No other science is concerned with the occurrence of rocks in space, and so geology where necessary has developed systems of its own.

Landforms, rock structures, rock formations, and a full range
of phenomena from earthquakes to weathering are described
and classified in purely geologic terms. There is also an unbelievably complex nomenclature for all these rocks, minerals,
landforms, and other geologic features; it is used by the pro-
fessionals as a form of shorthand that makes communication
more efficient than it would be if each item had to be described
in full. For example, the word "granite" identifies a particular
type of rock. Everyone except a geologist knows exactly what
it is. The problem is the definition, which is partly a description
of the rock and partly a conclusion about its origin. The description is standard enough, but there are so many possible
origins for this particular composition that the older idea that
granite is the result of a long, slow freezing process from a
silicate melt of proper composition cannot be defended in all
cases. Here is the definition: "Technically, the term "granite"
is reserved for those granular quartz-bearing igneous rocks that
have potassium feldspar as the chief mineral."[1] The word "igneous," meaning "formed by solidification from a molten or partly
molten state,"[2] is the controversial term. When a geologist is
faced with classifying a specific rock in a specific field context
as granite, he must be sure he can prove it has had an igneous
history rather than a metamorphic rock history. Rocks that
closely resemble true granite may be produced by recrystallization of sediments at temperatures below the melting point necessary for an igneous classification. The definition of granite
is a typical example of the pitfalls of scientific terminology.
Such words must either be used correctly in technical communication or avoided completely. In the latter case, ideas must
be expressed in common, descriptive ways with the understanding that the concern is to communicate in a useful, if imprecise,
manner.

The material aspect also establishes a conclusion in the mind
of the observer. Once he sees a crystalline rock such as granite,
he thinks immediately about origin rather than about specific
composition. Identification of grain size, percent quartz, and other
minerals become the clues that lead him directly to conclusions.

3. *Chronologic relationships*

We have already devoted an entire chapter to these ideas,
so it is unnecessary to expand upon them. The chronologic
relationships comprise the various "wasnesses of the is," and

78

The special character-
istics of geology an an
historical science

as we have seen by examining the grain in the block of wood
(Fig. 6), they too are conclusions.

4. *Genetic relationships*

These are the standard conclusions obtained via the principle
of least astonishment. The four aspects are bound together in
the now familiar format:

$$\text{Spatial} + \text{Material} + \text{Chronologic} \longrightarrow \begin{pmatrix} \text{Imply via least} \\ \text{astonishment} \end{pmatrix} \longrightarrow \text{Genetic}$$

Professor Grope likens them to Alexander Dumas' musketeers
—Athos, Porthos, and Aramis—who do all the work. But it is the
dashing D'Artagnan, the genetic aspect, who makes the whole
venture worthwhile. Taken singly, the aspects have little meaning,
but together they form wonder stories! As we shall see, "D'Artagnan
a le joie de vivre!"

The value of the four-part method of looking at geology becomes readily apparent as its application changes the tourist's
response to the Grand Canyon in Arizona from incredulous wonder

Fig. 10. The Grand Canyon.

to thrilled comprehension. There is no other gorge in the world comparable to this. We shall consider first how and why such a feature exists, and then we shall approach the Grand Canyon as a window into the structure of the continent. We shall use all four critical aspects in our exploration. It would be redundant to identify them by name each time they are used. Descriptions of rocks are obviously part of the material aspect, and the facts of elevation are obviously part of the spatial aspect. The method of wearing away the rocks is part of the how and why statement of the genetic aspect, while ages in years or in relative states are parts of the chronologic aspect. Each aspect is easily identified, and the reasoning that connects them should be clear enough to be acceptable as common logic.

A gorge the size of the Grand Canyon can exist only in a vast, high, arid plateau crossed by a single large stream. The stream must be fed from higher sources outside the plateau and must discharge into the sea.

Those two sentences contain a brief but complete statement of the genetic aspect. The proof will require a good deal of supporting information, beginning with a general description of the field situation (Fig. 10).

The Grand Canyon is 5,000 feet deep and 250 miles long. The upper part of the river has another 2,000 feet to cut before approaching sea level. Only a vast, high plateau can contain such a gorge. The Colorado Plateau Province covers about 150,000 square miles in the states of Colorado, New Mexico, Arizona, and Utah. Thick layers of sedimentary rock spread out in nearly horizontal sheets under much of the area. The climate tends to be arid, supporting very few rivers that flow the year round. Gully-washing rains combined with wind erosion have stripped off many of the less-resistant layers, exposing surfaces called strip plains that are protected by the more resistant rocks. These plains range in elevation from 4,000 to 11,000 feet above sea level. It is into this structural block that the Colorado River has cut its gorge.

If the region were less arid, there would be many tributaries to the Colorado supported by local runoff. These would soon cut up the plateau to such a degree that the canyon would lose its character as a well-defined narrow gorge. Therefore the regional aridity is responsible for the preservation of the canyon walls and the uniqueness of the phenomenon. Since the Colorado River is not formed from local runoff, the water must come from an exterior

80

*The special character-
istics of geology as an
historical science*

source. That source must be higher than the plateau, for water must run downhill. Melted snows and rains from the high western side of the Rocky Mountains supply the Colorado with a very modest mean annual discharge of 17,000 cubic feet per second. The Arkansas River, which drains the more humid eastern side of the Rockies at the same general latitude and receives additional water from tributaries, particularly in its lower course, discharges 42,000 cubic feet per second. The Mississippi River has a mean annual discharge of 630,000 cubic feet per second but no Grand Canyon! The Colorado has a small discharge, but after looking at the gorge, one must join Mercutio in commenting "'Tis enough. 'Twil do."

Two aspects of the statement remain to be explained: the cutting of the gorge and the necessity for discharge into the sea. They have been saved for last because they are interrelated. Well over 2,000 cubic miles of rock are missing from the main canyon area. In addition, the bed load transported by the river is equivalent to the amount of sand and gravel that could be carried by a continuous railroad train moving loaded gondolas along at a rate of three to four miles per hour. All of this has been going on for about 10 million years. Only the sea, in this case the Gulf of Lower California, is large enough to hold the accumulation. Abrasion of the bed and downcutting are accomplished more by the action of moving sediment than by water alone. Bottom scour probably occurs most actively during great floods, when velocities and transporting powers are increased many fold. The discovery of a wooden board with saw marks on it at a depth of 80 feet in the riverbed gravels when the foundation was dug for Hoover Dam is a dramatic illustration of this tremendous scouring power.

To appreciate the Grand Canyon, think of the river as a stranger to the high, arid plateau. Most of the erosion has occurred in the narrow floor of the river itself, for this is the only place the stranger has visited. The active zone of erosion is only about one fifth of a mile wide. Once the river has cut below each successive horizon, the walls weather back at an angle of repose that is dependent upon the materials. Strong rocks stand in nearly vertical cliffs; weaker rocks have gentler slopes. The tributaries that do flow year round have cut deep, narrow valleys that meet the Colorado on steep courses of their own. Between these valleys are the walls that are composed of rocks that give us a glimpse of the structure of a typical continental plate. These walls are a window into the past.

It is on the wall rocks that geologists tend to focus most of their attention.

We shall now prove that the core of the continent is very ancient. It has had a long history of its own, during which it was deeply eroded. At least 12 vertical miles of rock that composed the original regional overburden were removed before the area was covered again by thousands of feet of marine and continental sedimentary strata.

Looking back at Fig. 10, the data are all there if we identify details. The rocks of the inner gorge (Ig) are obviously different from those of the layered units above. Referring to the material aspect, we find that the rocks of the inner gorge are a complicated series of gneisses, schists, granites, and marbles. The minerals that compose these rocks are primarily quartz, feldspar, mica, pyribole, and where marbles are present, calcite and dolomite. The first group are all compounds of silicon and oxygen with other metals, including iron, calcium, magnesium, manganese, aluminum, sodium, and potassium. Combinations such as these are formed under pressures approaching 5 kilobars and temperatures of between 300° and 600° centigrade. Translated into less technical language, the rocks found in the inner gorge could *not* have been formed under the low pressure and temperature conditions present there today. During the formation of these silicate minerals, the overburden pressure must have been equivalent to that produced by a column of rock 12 miles high! All this had been eroded away prior to the deposition of the lowest sandstone (T).

The chronologic aspect helps us to understand what we see. The crystalline rocks of the inner gorge are about 1.5 billion years old. The first sedimentary layer above them is only 0.55 billion years old. This means that during an interval of about 0.9 billion years (900 or so million years), active erosion in this region removed 12 miles of rock on a regional scale. The surface left by this erosion occurs at the top of the crystalline rocks and just below the lowest sand grain at the bottom of layer T. It may be hard to picture this surface from the small sample seen in Fig. 10, where only the edge is exposed to view. Some idea of the regional nature of this erosion may be grasped in the way in which the surface has been cut by the side canyons. There are actually more data on the regional rock history, erosion, and movement than are given in this one picture. We are not attempting to be fully descriptive but to show how to truly see an old erosion surface

82

*The special character-
istics of geology as an
historical science*

on which the sun once shone, the clouds dropped their rain, and across which streams once flowed to all but forgotten seas.

There are many surfaces of profound erosion like this existing today. The best example in the United States is found halfway across the continent and is called the Piedmont Province of the southeastern states. It lies between the Coastal Plain Province and the Appalachian Mountains and stretches for hundreds of miles between southern Pennsylvania and central Georgia. The Piedmont rocks are also made of silicate minerals that were formed under great pressures and temperatures. They too have been exposed by erosion that has removed between 12 and 18 miles of overburden. The inhabitants of this region do not think of it as an erosional wonderland. They do not look up and marvel that the streams carried off 12 to 18 miles of rock before the surface on which they live was made. Without steep canyon walls, the awareness of total erosion depends upon an appreciation of the chemical boundary conditions necessary for the formation of the minerals in the rocks. Most of the general public do not have this type of specialized information. People are quite aware that each time it rains the streams run in muddy spate, but they do not notice the cumulative effects of erosion and transportation. For example, a stream that is only one foot deep and twenty feet wide, flowing at a rate of about two miles an hour, may contain two percent mud and other material in seaward-moving suspension. In one hour 200,000 cubic feet of water passes a given point on the bank. If we consider that two percent of this is solid material, its volume is 4,000 cubic feet. This means that such a stream transports enough mud in one hour to fill a box ten feet wide, ten feet high, and forty feet long. Obviously this stream and others like it would have little difficulty in removing a 12-mile overburden during a period of 350 million years.

Returning once more to the Grand Canyon, the line that separates the crystalline rocks of the inner gorge from the lowest sedimentary layer (T) may be seen in a fascinating new way. It is no longer a line. It is a wide stage. The streams that made it were drowned out long, long ago, not by the covering sandstone layers alone but by the waves and waters of advancing seas that deposited these sedimentary rocks. As the waves crept higher, they beveled the rocks even more with their own horizontal booming charges and, of course, smoothed out the sands that had been laid down as a coastal-plain complex of deltas, floodplains, and estuary fills. In this way the transition was made between the crystalline rocks with their own history and the sedimentary cover that rests on them. However, if we are to use the Grand Canyon as a window and peer through it for data that will let us generalize about the structure of the continent, we should have some assurance that this is a representative sample of the crystalline core.

The representative nature of the rocks of the inner gorge can be demonstrated by comparing them with rocks found in the southeastern Piedmont Province. This is a large area of similar rock types, similar internal structures, and similar erosional history that exhibits a similar final landscape form. Fortunately there are a large number of equivalent, although smaller, exposures in many parts of the United States. There is also a tremendous area of the Canadian Shield, extending from the Adirondack Mountains of New York State to the Arctic rim, that is composed of similar crystalline rocks. This same material composes the roots of great mountain chains throughout the world, including the Sierra Nevada, Andes, Himalayas, and Alps. They are familiar parts of the mountain root structure, and their presence proves that deeply eroded mountain roots are the basic continental building blocks. They may be likened to the individual tiles in a mosaic floor. We can therefore be confident that rocks exposed in the inner gorge of the Grand Canyon give us a view of the roots of mountains lost long before the lowest sediments were deposited.

We must now move up to the sedimentary section, the layers labeled T through K in Fig. 10. Sandstones, limestones, and shales comprise the upper 4,000 feet of the canyon wall. Sandstones are made of small grains of quartz, each one a record of a previous history. They are tiny rounded pellets that, when seen under the microscope, show signs of the abrasion resulting from the tumbling contacts that occurred in wind and water currents as they were being transported from older source rocks to their final resting places. Sandstones such as those of strata T and C serve as temporary storage places between cycles of erosion, transportation, and redeposition. Quartz is very durable. It is hard, chemically inert, and almost insoluble in water. Grains that were deposited in layer T some 550 million years ago were released as the Colorado River cut the canyon. They now lie in the Gulf of Lower California, with a mixture of similar grains from other sources, as part of a new rock in a new cycle. All of the quartz grains originate in crystalline rocks, and after being loosened by weathering,

84

*The special character-
istics of geology as an
historical science*

they begin the long, roving history that lasts until the rocks they
form are melted and become new granites again. The cycle extends
far into the past. James Hutton is known as the father of modern
geology because he learned how to read the rock record. His book,
Theory of the Earth, was published in 1788. The last sentence of it
served to open the way for Sir Charles Lyell and Charles Darwin.
This is Hutton's grasp of time. *'The result therefore, of our present
inquiry is, that we find no vestige of a beginning, no prospect of an end*
[italics mine]."[3]

The shales for the most part are made of clay. This group of
minerals, for there are many types of clay, is produced by chemical
weathering that adds water to the molecules of some of the source
mineral forms found in the crystalline rocks. Minerals such as
feldspar, mica, and pyribole cannot exist in humid climates without being converted to other minerals. Clay and two iron oxides,
limonite and hematite, are commonly formed by the weathering.
Most of the colors ranging from red to golden brown that are seen
in the Grand Canyon are the result of iron-stained clays that were
originally deposited as marine shales and are now exposed in
terraces and minor cliffs. The large central unit (R), which forms
a cliff nearly 600 feet high, is a marine limestone stained from
grays to reds by the shales weathering away above it. The highest
layer (K) is also a marine limestone, but it remains white for there
is nothing above left to stain it. Layer C, on which the observer
stands on the near rim of the canyon in Fig. 10, is a white sandstone that was originally formed for the most part by the action
of desert winds. Cross-bedded structures visible on the outcrops
are a record of otherwise forgotten winds that blew across this
region about 260 million years ago.

The entire 4,000-foot section may be taken apart grain by grain,
fossil by fossil, and its story related as part of the picture-window
view. Layer T, at the base, contains 550 million-year-old trilobite
fossils from the middle Cambrian period. Evolution had proceeded
on the earth at such a pace that some vertebrates had already
achieved air-breathing status before the cliff-forming rocks in layer
R were deposited. Giant amphibians as large as half-grown alli-
gators have been found as fossils in the layer between C and K.
Evolution of the higher forms, first of dinosaurs and eventually
mammals, including of course man, occurred after these Grand
Canyon layers were laid down. Nevertheless the picture window
does provide a good view of the early stages of animal evolution.

Layers C and K were deposited in the Permian period. We had some experience with the Permian period in Chapter 2. This was the time of the glaciation of South Africa, Australia, and India. The Grand Canyon section does not show any glacial rocks, and we can be sure that the area was not in a part of the earth cold enough to be covered with ice. The amphibian fossils are good evidence of the climatic ranges during their evolution. Amphibians need to spend a good deal of time in the water and reproduce through an aquatic cycle in the same manner as frogs do today, passing through egg and tadpole to adult stages.

The Permian was the last period in which this area lay beneath the sea. Therefore, for some 200 or more million years, erosion either has been occurring or has been impeded by continental deposition. The Grand Canyon itself is quite young, so there is a complicated story of other rivers and other activities that could be told about the region prior to the time of the cutting of the deep gorge that has given us this window into the past.

One of the most enjoyable aspects of this glimpse into the structure of the continent is the recognition of changes in the position of things. Even the floor of the inner gorge is 2,000 feet above the sea today. The gorge is about 1,200 feet deep. Therefore the first marine unit (T) is found today at an elevation of about 3,500 feet above sea level, and the highest marine layer (K) is more than 8,000 feet above sea level. We can see that the continent has been uplifted, depressed, and uplifted again quite a few times. It would be illogical to imagine that the oceans moved up and down this much, for on a worldwide basis such movement would require the gain and loss of tremendous volumes of water. Adherence to the first principle of science requires us to rule out the idea that water is made and lost in some miraculous fashion.

Lyell pointed out the phenomenon of the uplift and depression of land in respect to sea level in volume II of his *Principles of Geology*.[4] A picture of the Temple of Serapis at Pozzuoli, Italy, on the north rim of the Bay of Naples, appeared as the frontispiece of volume I in all the early editions of this book, and it has been reproduced here as Fig. 11. The temple was built by the Romans not more than a few centuries before the Christian era. It has since been depressed beneath the sea and then lifted again to its present position. The vertical movement in each case was between 20 and 25 feet. The columns are 42 feet high, and deep scars produced by the boring marine mollusk Lithodomus cover part of the lower

86

*The special character-
istics of geology as an
historical science*

Former sea
level marked
by scars of
boring mollusk
Lithodomus

Fig. 11. Temple of Serapis, Pozzuoli, Italy. Sir Charles Lyell used the data provided by the columns to demonstrate both elevation and subsidence of the
land. (From Lyell, Sir Charles. 1842. Principles of geology. Vol. I. Hilliard &
Gray & Co., Boston.)

surfaces of each column. When the pillars were depressed below sea level, mollusks made their homes by drilling into the marble. Lyell pointed out that the movements were on a local scale and did not affect large regions. This is proof enough that the land moved with respect to the sea as opposed to the sea level rising and falling while the land remained stable. Changes in sea level do occur. After the glaciers of the last million years melted away, the level of the sea rose by between 325 and 470 feet around the world. The data found in the Grand Canyon, however, do not represent worldwide sea-level changes. The land came up, and with the increase in height, the stream gradients increased. As water speed increased, so did the cutting power of the rivers. The entrenchment at the Grand Canyon is certainly the most remarkable of its kind on earth. Even so, the total result is exactly what should be expected.

Our new insight into regional uplift and depression has made it possible to employ another of the four critical aspects. It is impossible to describe the position of the Temple of Serapis properly without also indicating a time. The land moves up and down, and so its position in space becomes a matter of where and when. Similarly, central Arizona is above the sea today, but some 240 million years ago it was beneath the sea as layer K was deposited. Layer C, below layer K, is a continental deposit, and the area was above sea level when it was formed, perhaps 260 million years ago.

Geologists view the four critical aspects using the "where-when," "where-what," and "what-when" question pairs. Each of these pairs becomes a conclusion formed so naturally that it seems to be a single fact. Geologists peering into the Grand Canyon think in terms of older and younger rather than bottom and top or down and up. Historians do exactly the same thing when they discuss the human activities that took place at some time in the past. Their point of reference is the character and quality of that age rather than their own. This is all geologists do as they restore the sunlight, rain, and streams to the ancient erosion surface at the top of the inner gorge.

There are two great keys to enjoying geology. One of them is the ability to think along the lines described in the last two chapters. The other involves retaining contact with the world around you, wherever you are. Stressing technical ideas is not enough. There is another side of geology as it relates to man. Let's not leave the Grand Canyon until we have seen that, too!

88

*The special character-
istics of geology as an
historical science*

Spanish conquistadors discovered the Grand Canyon in 1540,
only 48 years after Columbus landed in the Bahamas. Coronado
was searching for the fabled Seven Cities of Cibola and one of
his captains, Don Garcia Lopez de Cadensa, stumbled onto the
canyon's rim. A few days were spent fruitlessly trying to reach
the bottom. Lopez de Cadensa returned to the top with the comment that rocks that looked as small as a man from above were
"bigger than the great tower of Seville."[5]

Three hundred and nineteen years later, the canyon was still a
mystery. No one knew where it led or what cataracts it might
contain until Major John Wesley Powell led an expedition through
it by boat in 1869. Powell staked the lives of his entire party on
the belief that a river with a gradient the magnitude of the Colorado's would not contain waterfalls like Niagara. He was right.
The erosion had been so effective that there were only great rapids
that could either be negotiated or bypassed. Powell had his boats
made of oak in Michigan and shipped by railroad to the Green
River crossing in Wyoming. His party left civilization on this tributary of the Colorado River just ten days after the golden spike
had been driven at Promontory, Utah, to mark the spanning of
the continent with parallel strips of steel set on wooden crossties
four feet, eight inches apart. The new era of economic development began just as Powell's three-month adventure brought the
era of the great North American explorations to a close. This was
the end of the unknown.

A century later, in 1969, there were nineteen commercial float
trips taking tourists through the same rapids that Powell had found
concentrated at the mouths of the few tributary streams. The occasional desert rains brought great boulders down the steep gradients of the tributaries, and even the power of the Colorado River
required time to wear them away. The seething violence of the
water at these points is the principal danger to boats and men.

Geology is an outdoor science. Even the laboratory work is
based on materials and data gathered in the field. Explorers like
Major Powell have played a great part in gathering the information and determining its significance. Yet there is another aspect
of outdoor geologic activity—helping students to discover the language for themselves. The Reverend John Walker, who taught the
first systematic geology courses in the English language at the
University of Edinburgh in Scotland from 1781 to 1803, told his
students to get out in the field in these words:

The objects of nature themselves must be sedulously examined in their native state, the fields and the mountains must be traversed, the woods and the waters must be explored, the ocean must be fathomed and its shores scrutinized by everyone that would become proficient in natural knowledge. . . . The way to knowledge of natural history is to go to the fields, the mountains, the oceans, and to observe, collect, identify, experiment and study.[6]

Professor Grope is so taken with Walker's philosophy that he has structured his own classes on the same principle. Every Wednesday afternoon Grope and his students can be found in the field. This exposes them to a good deal of healthy sunshine, some wind, rain, and occasional snow. Grope contends that the elements are agents of geologic change that students should feel privileged to view at first hand. His bravado does not fool anyone, but it makes light of some bad situations. Where outcrops lie some distance from the roads and his classes are forced to walk, the rear guard may be heard grumbling heartily, as though they were the last stragglers in Xenophon's column. Comments about hiking, merit badges, and the wisdom of enlisting in geology next semester are typical. Once an English major was caught with Grope in a spring shower, and standing drenched to his socks, he mused about "the uncertain glory of an April day." Grope has chuckled over that one for years. John Walker, despite the formality of instruction in those days, must have enjoyed similar outings and similar comments from the company!

Contrast all of this with the approach of the rock hounds, who think that geology is collecting things. Grope is seldom struck dumb, but rock hounds leave him speechless. One group invited him to their sixth annual rock swap with the suggestion that he might be able to pass off some bad rocks for some good ones. The musketeers are of little value unless they function as a team.

ANNOTATED REFERENCES

1. Gilluly, James, Waters, Aaron C., and Woodford, A. O. 1968. Principles of geology (ed. 3). W. H. Freeman & Co., San Francisco. 678 pp. (The definition of granite is taken from p. 613. Used by permission of W. H. Freeman & Co.)

2. Towbridge, A. C. (ed.). 1962. Dictionary of geologic terms. Dolphin Books, Doubleday & Co., Inc., Garden City, N. Y. 645 pp.

3. Hutton, James. 1970. Theory of the earth. Hafner Publishing Co., Darien, Conn. 203 pp. (This is a facsimile of the 1788 edition. The quotation is found on p. 304 in the original system of numbering pages.)

4. Lyell, Sir Charles. 1842. Principles of geology. Hilliard & Gray & Co., Boston. Vol. I, 442 pp. Vol. II, 477 pp. (The frontispiece of volume I, showing the Temple of Serapis at Pozzuoli, Italy [see Fig. 11], is discussed in volume II, p. 384 ff. The three volumes of the first edition were published in 1830, 1832, and 1833, respectively. I have not examined this rare book personally. The third English edition, published in 1834, contained the same material on the Temple of Serapis but in a little less detail.)

5. Lowery, Woodbury. 1959. The Spanish settlements within the present limits of the United States. 1513-1561. Russell & Russell Publishers, New York, 515 pp. (The remark by Lopez de Cadensa is quoted from p. 311.)

6. Scott, Harold W. 1966. Lectures on geology by John Walker. University of Chicago Press, Chicago. 280 pp. (Walker's admonishment is taken from the preface, p. xvii. Used by permission of the University of Chicago Press.)

PART IV

Illustrations showing how you use all of these concepts "to see a world!"

Chapter 7

There's a legion that never was 'listed,
　That carries no colours or crest.
But split in a thousand detachments,
　Is breaking the road for the rest.

The Lost Legion, Rudyard Kipling

The discovery of geology in ancient times

We are now armed intellectually to join Herodotus in Egypt sometime between 460 and 443 B.C. to make a geologic survey of the country under his direction. Herodotus was a geologic genius. His logical approach to the analysis of field data would have done credit to James Hutton, Charles Lyell, or Phil King. Herodotus will conduct the class himself from his own text. The presentation is taken out of context and rearranged to the extent that extraneous material is deleted. In order to avoid confusing footnotes the text is irreverently interrupted to place necessary explanations where they will be most helpful. No attempt will be made to explain Herodotus' references to authorities of his day or to define his status as an historian. Our main concern is with the manner in which the first great regional geologist managed to draw such magnificent conclusions from the raw data he saw around him.

And they [Egyptian priests] told me that the first man who ruled over Egypt was Men, and that in his time all Egypt, except the Thebaic canton [area near Thebes about 400 miles inland; Fig. 12], was a marsh, none of the land below Lake Moeris [modern Birket Qarum, about 150 miles inland and above Cairo], then showing itself above the surface of the water. This is a distance of seven days' sail from the **93**

*Illustrations showing
how you use all of
these concepts*

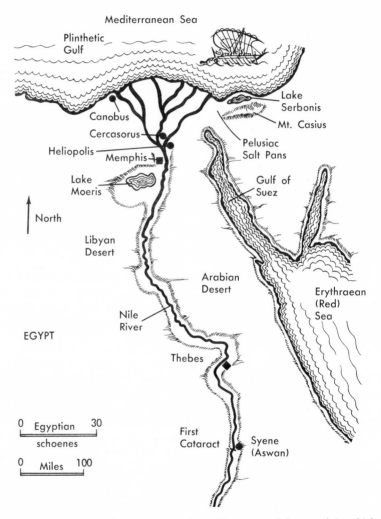

Fig. 12. Egypt as it was known to Herodotus. He compared the trough in which the Nile River flows with the Gulf of Suez and the adjacent portion of the Red Sea. He then suggested that the Nile could fill the latter troughlike areas with sediment in 10,000 to 20,000 years. (From Kiepert, Henry. 1902. Atlas antiquitus, twelve maps of the ancient world for schools and colleges, ed. 11. Leach Shewell & Sanborn, Boston.)

sea up the river. [This much may be thought of as instruction in geology by the Egyptian priests. In the next passage Herodotus' own genius begins to emerge!]

What they said of their country seemed to me to be very reasonable. For anyone who sees Egypt, without having heard a word about it before, must perceive, *if he has only common powers of observation* [italics mine], that the Egypt to which the Greeks go in their ships is an acquired country, the gift of the river. The same is true of the land above the lake, to the distance of three days' voyage, concerning which the Egyptians say nothing, but is exactly the same kind of country.

The following is the general character of the region. In the first place, on approaching it by sea, *when you are still a day's sail from the land, if you let down a sounding-line you will bring up mud, and find yourself in eleven fathoms [of] water, which shows that the soil washed down by the stream extends to that distance* [italics mine.] [Herodotus must have been curious enough to do it. The passage also implies that he had a general knowledge of the sea floor and thought that unsorted, clay-rich river muds were distinctly unusual.]

From the coast inland as far as Heliopolis [very close to modern Cairo and at the head of the distributaries], the breadth of Egypt is considerable, the country is flat, without springs, and full of swamps. As one proceeds beyond Heliopolis up the country, Egypt becomes narrow, the Arabian range of hills which has a direction from north to south, shutting it in upon the one side, and the Lybian range on the other. The former ridge runs on without a break, and stretches away to the sea called the Erythraean [Red Sea]; it contains the quarries whence the stone was cut for the pyramids of Memphis [Cheops and others]: and this is the point where it ceases its first direction, and bends away in the manner above indicated.

On the Lybian side, the other ridge whereon the pyramids stand, is rocky and covered with sand; its direction is the same as that of the Arabian ridge in the first part of its course. Above Heliopolis then, there is no great breadth of territory for such a country as Egypt, but during four days' sail Egypt is narrow; the valley between the two ranges is a level plain, and seemed to me to be, at the narrowest point, not more than 200 furlongs [about 25 miles] across from the Arabian to the Lybian hills. Above this point Egypt again widens. [Herodotus has drawn a word-picture map, using the principle of adjacency to tie the parts together.]

The greater portion of the country above described seemed to me to be, as the priests declared, a tract gained by the inhabitants. For the whole region above Memphis, lying between the two ranges of hills that have been spoken of, appeared evidently to have formed at one time a gulf of the sea. It resembles to compare small things

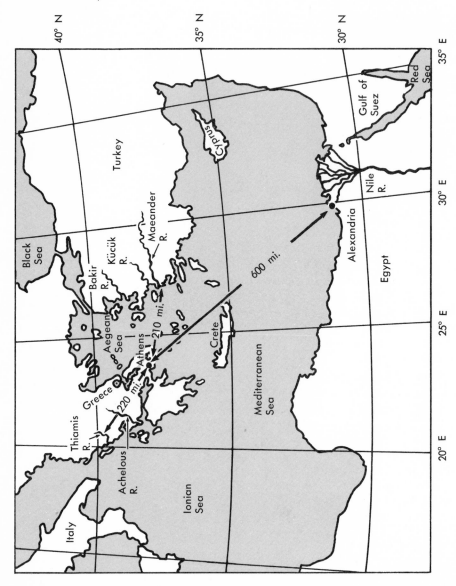

Fig. 13. The rivers cited by Herodotus as field evidence for his geologic conclusions.

with great, the parts about Ilium, and Teuthrania, Ephesus, and the plain of the Maeander. In all of these regions the land has been formed by rivers, whereof the greatest is not to be compared to any of the five mouths of the Nile. I could mention other rivers also, far inferior to the Nile in magnitude, that have effected great changes. Among these not the least is the Achelous, which after passing through Acarnania, empties itself into the sea opposite the islands Echinades, and has already joined one-half of them to the continent.[1]

The range of field examples available to Herodotus is extremely interesting. He was familiar enough with the general character of river action to cite two examples in Greece some 200 miles northwest of Athens and three examples in modern Turkey, about 200 miles east of Athens (Fig. 13), of features similar to those of the Nile 600 miles to the southeast. The Ilium area (Fig. 14, *A*) is on the west coast of Greece about ten miles from the Albanian border. The flat depositional plain of the Thiamus River is readily distinguishable from the surrounding hills. The long point is the delta of the river as it is being built into the Ionian Sea. Teuthrania is on the mainland of modern Turkey (Fig. 14, *B*) just east of the island of Lesbos. The delta that is being built into the Aegean shelf has been modified by wave action, which has produced the rounded form. The Ephesus area (Fig. 14, *C*) is also on the coast of Turkey, just northeast of the island of Samos. The Kücük River flows across a narrow depositional plain into the Aegean Sea. The Maeander River flows across a wide depositional plain between steep hilly walls. The winding contortions of the Maeander are easily seen on the map. Our use of the term "meandering stream" is derived from the name and nature of this river. The Achelous River (Fig. 14, *D*) is in central Greece at the mouth of the Gulf of Patras, where it opens into the Ionian Sea. The delta of the Achelous has obviously been built out from the land and has connected the series of isolated islands together, so that they now appear to be hills on the delta. The four maps in Fig. 14 were redrawn from a series used by the German army in World War II. The details known to Herodotus are fully evident today. Through these maps we can share his cone of vision.

One of the wise teachers who introduced Professor Grope to geology insisted that the outdoor science could be learned in only one place—in the field. He felt that a student could learn the same physics or chemistry in New York, Tokyo, Berlin, or Moscow but that geology was taught in terms of the local field exposures and

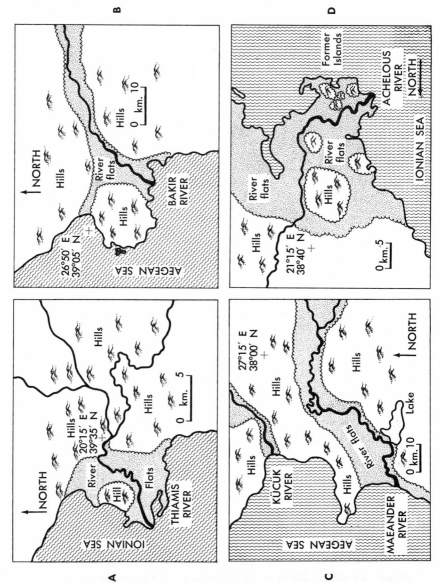

Fig. 14. More detailed maps of the river plains and deltas in modern Greece and Turkey. Herodotus saw that these were depositional phenomena comparable to the Nile.

the experiences that each faculty member had, as he said, "out yonder!" For this reason he recommended that students work for the United States Geological Survey as a means of accumulating experience in many areas before "pretending" to practice geology as a profession. This aspect of the science does not change. Herodotus had become familiar with the geology of Greece and Turkey before he saw that of Egypt. His amazing ability to generalize about the significance of similarities between the river plains and deltas in so many places is the key to his competence as a field geologist.

> In Arabia, not far from Egypt, there is a long and narrow gulf running inland from the sea called the Erythrean, of which I will here set down the dimensions. Starting from its innermost recess, and using a rowboat, you take forty days to reach the open main, while you may cross the gulf at its widest part in the space of half a day. In this sea there is an ebb and flow of the tide every day.[1]

At this point it is very important to determine the exact place (Fig. 12) Herodotus had in mind. The data are a little contradictory. *Webster's Third International Dictionary* assures us that the Erythrean Sea is the Red Sea, citing the Greek root word for "red" as evidence. Other authorities differ and identify it as the Arabian Sea, but that does not fit the description. The real problem concerns the dimensions. The Red Sea is between 100 and 200 miles wide and 1,200 to 1,300 miles long. No one could row across in half a day. The Gulf of Suez is only 30 miles wide, but it is 180 to 200 miles long. We shall assume that Herodotus was confused by hearsay evidence and meant the Gulf of Suez and some part of the northern end of the Red Sea in his next comments. This is an attractive idea, for the dimensions of the landforms are quite similar to those of the Nile valley and the more open delta country he has already described. The comparison of the Nile valley and the Gulf of Suez —Red Sea area, as shown in Fig. 12, is quite apt.

> My opinion is, that Egypt was formerly very much such a gulf as this—one gulf penetrated from the sea that washes Egypt on the north, and extended itself towards Ethiopia; another entered from the southern ocean, and stretched towards Syria; the two gulfs ran into the land so as to almost meet each other, and left between them only a very narrow tract of country. Now if the Nile should choose to divert his waters from their present bed into this Arabian gulf, what is there to hinder it from being filled up by the stream within, at the utmost, twenty thousand years? For my part, I think it would be filled in half the time. How then should not a gulf, even of much greater

size, have been filled up in the ages that passed before I was born, by a river that is at once so large and so given to working changes?[1]

In a logical sense Herodotus is guilty of begging the question. Nevertheless his grasp of the fact of change through time and of the reality of long spans of past time is truly remarkable. The Nile valley has been cut into the higher plateaus on either side by the erosion of the river during periods of much lower sea level. In the glacial ages that began about 2 million years ago and ended less than 20,000 years ago, the sea level was lowered by several hundred feet as water was accumulated in the extensive ice sheets. All of the major coastal rivers were able to cut more deeply at that time. Most of the fine harbors of the world are flooded river valleys. Actually two causes work together to account for them. Flooding represents a high sea level due to the melting of Pleistocene ice sheets. Harbors rather than deltas at the coastline are evidence that the rivers do not supply enough sediment to fill the old valleys. The Nile and the Mississippi have filled their well-marked valleys for hundreds of miles upstream since glacial times. On the other hand, the Hudson has not filled New York harbor since the flooding. Neither have the San Joaquin or Sacramento Rivers filled San Francisco Bay. Chesapeake Bay is the flooded mouth of the Susquehanna River. Hampton Roads is a part of the Chesapeake Bay and also the flooded mouth of the James River. The first ironclads, the *Monitor* and the *Merrimac,* battled here during the hostilities of 1862. So we see that Herodotus was not really very far off in his estimate of the amount of time that large, silt-laden rivers would require to fill major estuaries and valleys.

The Gulf of Aden, Red Sea, Gulf of Suez, Gulf of Aqaba, and the lowland trench of the Holy Land are remarkable, troughlike structures bounded by steep walls of broken rock. These breaks, or faults, were formed as Africa and Arabia moved apart in the process of continental drift. Frequent earthquakes that rock the southern part of the Red Sea today are testimony that the process is still in action. The floors of all the troughlike structures have been dropped below sea level in much the same manner as a roof would fall if the walls of a building were moved apart. The Sea of Galilee, Jordan River, and Dead Sea, the latter with a surface 1,286 feet below sea level, are all bodies of water confined in the same trough that holds the Gulf of Aqaba. Herodotus compared the external aspects of the lowland regions of the Nile Valley and Gulf of Suez, not their internal aspects. His real intellectual breakthrough consisted in bringing a regional situation into focus in his cone of vision and considering change through time in a rational manner.

Thus I give credit to those from whom I received this account of Egypt, and am myself, moreover, strongly of the same opinion, since I remarked that the country projects into the sea further than the neighboring shores, and I observed there were shells on the hills, and that salt extruded from the soil to such an extent as even to injure the pyramids. . . .[1]

These famous lines are often quoted as proof that Herodotus was the first paleontologist. He saw shells in the rocks and he called them shells. This is a bold, straightforward approach. Several Chinese philosophers of the twelfth century A.D. made similar observations and drew similar conclusions. It would seem that from such an early, universal beginning, the science of paleontology should have been highly developed long before the nineteenth century. The reason for the long delay was the position taken by many religious faiths in the controversy involving interpretation of time and biblical interpretations of the Noachian flood. Charles Gillespie, in his book *Genesis and Geology,*[2] describes some of the ramifications of the whole range of geologic discoveries in the 80-year period just prior to Darwin. Apparently the fossil controversy raged for centuries because of the insistence that biblical accounts precluded the possibility of ancient fossil life. The least astonishment conclusions based on rock evidence could not be squared with literal interpretations of the Bible. The impasse was finally broken when so much data had been accumulated that fossils could no longer be ignored as insignificant accidents. Nineteenth-century geology moved forward with vigor as fossils, which had served to trigger the understanding of organic evolution, became also the tools for correlating rocks in time and for interpreting the environments of deposition. This pattern of change is a typical illustration of the crisis of contradiction. Herodotus was not forced to struggle with dogma and he made good use of his data.

Herodotus' lines are also startling in their recognition that, because the Nile Delta protrudes into the Mediterranean, the bulge is a record of the extension of the delta by deposition. If Herodotus had sailed to Egypt by navigating the eastern coastal waters of the Mediterranean, he would have had to change course to pass around the curve of the delta before he could have reached the western distributary mouth near modern Alexandria. We do not know that he entered Egypt through the western rather than through a more easterly distributary, but obviously he must somehow have been aware of the delta form. The average tourist would probably not consider that a shoreline shape had special meaning. Herodotus had already seen the same effect in the delta of the Achelous River, which protruded and tied half of the Echinide Islands to the mainland (Fig. 14, D). The delta of the Thiamis River (Fig. 14, A) protrudes in a similar way. These are further examples of the manner in which Herodotus used specific observations to draw generalized, widely applicable conclusions.

> ... and I noticed that there was but a single hill in all of Egypt [speaking of the confined Nile valley] where sand is found, namely, the hill above Memphis; and further, I found the country to bear no resemblance either to its borderland Arabia, or to Lybia—nay, nor even to Syria, which forms the seaboard of Arabia; but whereas the soil of Lybia is, we know, sandy and of a reddish hue, and that of Arabia and Syria inclines to stone and clay, Egypt has a soil that is black and crumbly, as being alluvial and formed of the deposits brought down by the river from Ethiopia.
>
> One fact which I learnt of the priests is to me strong evidence of the origin of the country. They said that when Moeris was king, the Nile overflowed all Egypt below Memphis, as soon as it rose so little as eight cubits [12 feet]. Now Moeris had not been dead 900 years at the time when I heard this of the priests; yet at the present day, unless the river rises sixteen, or, at least fifteen cubits [22.5 to 24 feet], it does not overflow the lands. It seems to me, therefore, that if the land goes on rising and growing at this rate, the Egyptians who dwell below Lake Moeris, in the delta, as it is called, and elsewhere, will one day, by stoppage of the inundations, suffer permanently the fate which they told me they expected would some time or other befall the Greeks [famine by drought].[1]

Herodotus speaks of time with the ready familiarity of a scholar who has lived with the wasness of the is. His reference to an event 900 years past is comparable to one of our contemporaries saying, "The Norman conquest had taken place only 900 years before I heard about it." Herodotus' discussion of the rate of deposition is difficult to evaluate. The Egyptians did have Nilometers[3] to measure the rise and fall of the river. Some of them were more than 1,000 years old in Herodotus' time, so that the priests could have had valid measurements on which to base their information.

At the least his statements are a recognition that past change is a reality and that in the future further change may be anticipated.

If then we choose to adopt the views of the Ionians concerning Egypt, we must come to the conclusion that the Egyptians had formerly no country at all. For the Ionians say that nothing is really Egypt [but] the Delta, which extends along the shore from the Watch-tower of Perseus [?], as it is called, to the Pelusaic Salt-pans [just east of modern Port Said], a distance of forty schoenes [about 300 miles by Herodotus' conversion factors], and stretches inland as far as the city of Cercasorus [near modern Cairo], where the Nile divides into two streams which reach the sea at Pelusium and Canobus [near modern Alexandria] respectively. The rest of what is accounted Egypt belongs, they say, either to Arabia or Lybia. But the Delta, as the Egyptians affirm, and as I myself am persuaded, is formed of the deposits of the river, *and has only recently, if I may use the expression, come to light* [italics mine]. If, then, they had formerly no territory at all, how came they to be so extravagant as to fancy themselves the most ancient race in the world? . . . But in truth I do not believe that the Egyptians came into being at the same time with the Delta, as the Ionians call it; I think they have always existed ever since the human race began; as the land went on increasing, part of the population came down into the new country, part remained in their old settlements. . . .¹

This passage is so clear that comments on it are almost redundant. Herodotus has complete appreciation of the four critical aspects: material, spatial, chronologic, and genetic. A grasp of time has appeared repeatedly in the account, but never as startlingly as in his apology for using the word "recently." Acceptance of the antiquity of the human race is a concept that would have been strange to many well-educated individuals as late as the nineteenth century. Dunbar⁴ gives a number of the estimates of the age of the earth prepared by biblical scholars on the basis of the "begats" and other things. These range from 9:00 A.M. on September 17, 3928 B.C. (by John Lightfoot, Vice Chancellor of Cambridge University, in two publications dated 1642 and 1644), to the "entrance of night preceding the twentythird day of October, in the year of the Julian Calendar, 710, i.e., 4004 B.C." (Archbishop Ussher, Primate of Ireland, 1658), to the simple date of 4004 B.C. inserted in the 1701 edition of the English Bible by Bishop Lloyd. The American poet Zerbino, writing in 1815, typifies this sort of time-scale thinking in *On the Passaic Falls*, a contemplation of the marvels of the rock record.

Amid these records of a deluged world,
By power divine, in awful ruins hurl'd;
Amid these works of patriarchal time,
Age after age, still beauteous, still sublime,
The poet loves to study nature's book,
And back four thousand years, in thought, to look. . . .[5]

Few people, even today, have any realistic conception of time
prior to their own first memories. For this we do draw on authorities, and it is not surprising to see the limitations imposed by
authorities who are in error. The wisdom of Herodotus is staggering
by contrast. How unfortunate that the libraries at Alexandria were
burned! Herodotus' vague reference to "the Delta, as the Ionians
call it," suggests that there may have been a teacher of the John
Walker variety before the sixth century B.C., predating Herodotus
by 200 years.

It is interesting to abstract from the writings of Herodotus the
underlying principles that shaped his perception and observation.
These may be called "the first, first principles of geology."[6]

1. *Herodotus acted on the basic presupposition of science—nature can be
understood.*

The prima facie case for this is the way he comprehended
all he saw. Had he not accepted this presupposition, he would
have been like most other travelers to Egypt before and since,
who have looked with unseeing eyes at the unfolding of the
terrain.

2. *Herodotus collected and classified field data with great care.*

An example of his method is his observation on the composition of the mud of the floor of the Mediterranean Sea.
Another example of classification and correlation is his citation
of the types of soils and rocks associated with different regions.
Modern geologic studies are based on similar although more
sophisticated kinds of correlation.

3. *Herodotus searched for the general or universal natural characteristics
that relate classes of phenomena; he was not content to accept a unique
explanation for each separate item.*

This is one of the most important aspects of his work. The
average man would not think of linking the appearance of river
valleys hundreds of miles away and many times less spectacular
in size with the valley of the great Nile. This is a remarkable
adaptation of the cone of vision concept, for Herodotus thought
of rivers as general phenomena into which the individual ex-

amples fitted as parts of a whole. His view of rivers is equally remarkable for his inclusion of the work they do in erosion, transportation, and deposition as a substantial part of the picture. People generally think of each river as unique; Herodotus thought of each river as a variation of a single form.

4. *Herodotus grasped the wasness of the is.*

Egypt was a good place in which to develop this concept, for by the time Herodotus arrived some of the monuments were already 2,500 years old. He could see time expressed as the duration between events. Even so, it is still remarkable that Herodotus was able to see a natural phenomenon like a river delta as an expression of action occurring through time in the same manner in which we examined the block of wood as a record of time.

5. *Herodotus understood that causality could be comprehended through a least astonishment approach.*

He used the standard format of a list of facts that implied a conclusion. Further proof of his grasp of the principle is the manner in which he singled out anomalous facts and restructured conclusions of his own rather than accept the conclusions of others. His discussion of the northward movement of the Egyptian people onto the delta as it increased in size with deposition is a good example of Herodotus' use of the least astonishment approach.

Taken together, all five of these generalizations constitute the first, first principles of geology. What would have to be added so that these principles would be applicable to modern science? Herodotus did not know the chemistry and physics necessary to study materials, and his classification procedures were limited. He did not know the full range of processes by which the uniformitarian analogy is used in least astonishment solutions. He did not know the biology necessary to make the most of fossil interpretations. Base maps for displaying the cone of vision concept in picture form had not been developed, so that his ability to compare regional data was more limited than his skill at making observations. Despite all these difficulties, the insight displayed by Herodotus may be explained in one word—genius.

The history of science is a wonderful study of false leads, blind alleys, dead ends, and all manner of intriguing insights. The pattern must be unraveled either on a de facto basis, as in the case of Stonehenge, or with the surviving literature, whatever it may be. If we look back to the time before Herodotus to see what might have been known about geology, there are scattered bits of information but no comprehensive description of a region. Literature is a useful source of information because it reveals the concepts generally understood by its audience, whatever the time or place. An author who used a geologic image to emphasize a point must have felt that the image would be understood; otherwise he would not have used it. We shall examine an example of such usage in the Old Testament. The setting is familiar enough. About 539 B.C. Cyrus permitted the Jews to return from Babylon to Jerusalem across the arid Arabian Peninsula. Isaiah is quoted in the Revised Standard Version:

A voice cries:
In the wilderness prepare the way of the Lord,
Make straight in the desert a highway for our God (40:3).

Every valley shall be lifted up,
and every mountain and hill shall be made low;
the uneven ground shall become level,
and the rough places a plain (40:4)

The King James Version is a little different:

Every valley shall be exalted,
and every mountain and hill shall be made low;
and the crooked shall be made straight
and the rough places a plain (40:4)

Luke quoting Isaiah in the New Testament is translated in both versions:

Every valley shall be filled,
every mountain and hill shall be brought low:
and the crooked shall be made straight,
and the rough places shall be made smooth (3:5).

This geologic imagery is not so evocative to those raised in humid regions where rivers flow to the sea, disposing of the sediment in an out-of-sight, out-of-mind manner. However, Isaiah and the folk intellectuals of his day, to whom his words had reasoned rather than magical meaning, were natives of an arid region with few permanent streams and totally different geologic actions.

Geologists became aware of this difference when they began to study the Great Basin (see Fig. 16) in the American West. Between the Rocky Mountains and the Sierra Nevada, in the western parts of Colorado and New Mexico, throughout most of Utah and Nevada, and in parts of Oregon, California, and Arizona, with an extension through the Sonoran and Chihuahuan deserts and south as far as Mexico City, the arid climate has produced an erosion system that would be more familiar to Isaiah than to residents of the Eastern Seaboard. Rain does fall and erode mountains, but because the amount of rainfall is slight and the streams lack power, debris tends to accumulate in the valleys, filling them up. The result of the completed arid cycle of erosion is that the hills are buried in their own debris, leaving a flattened topography due partly to erosion and partly to fill.

Some Western theologians present an exegesis for these verses that is tied to the boundary conditions of the humid cycle. The interpretation they place on them is that of road building. Bulldozers make a modern superhighway by cutting through the hills and filling the valleys to reduce the grades. The dictum of the civil engineer is to have the amount (volume) of cut balance the amount of fill. Roads made in this way tend to be fairly straight. It is amusing to imagine Isaiah using such modern imagery. A geologic historian might approach the analysis quite differently, avoiding exegesis entirely. He would not be concerned with the message that Isaiah meant to pass on to his people. The concern of the geologic historian is not what the literary image is used for, but what image is used? Taken at face value, the image is that of arid-cycle geology. We cannot know how much geology Isaiah knew, but he has obviously used facts that are familiar to desert people.

If the verses of Isaiah are a window to ancient geologic knowledge, they are not the only available source of information. Nelson Glueck spent many years in archeologic research in the Middle East. His reports[7,8] include evidence that the ancient peoples, responding to the requirements of dry farming, directed, collected, and stored water as flash floods and mountain rains occurred. Storm waters are heavily loaded with sands and silts. The cisterns and trenches must have been silted up repeatedly. There is little question that the ancients were well aware of the erosion, transportation, and deposition trilogy. It would be very natural for Isaiah to use it in his imagery.

Glueck also discusses some of the ancient roads.

Not having to consider automobile traffic, the Roman engineers could run their roads up and down the steep slopes of these great wudyan [valleys] in a much more direct line than the modern engineers, who are compelled to choose a much more zigzag course. And that is about the only main difference between these two roads, the one built near the beginning of our era and the other in the 20th century A.D. Other differences could be detailed, to be sure, but they would not cast a favorable light upon the modern roadbuilders in Trans-Jordan. The Romans paved their highway all the way from its very beginning to its very end. It was divided into two lanes, with a protruding line of stones in the middle, and the sides of the road also marked by raised stones. I have seen large stretches of the modern road washed away after the first rain. The Romans, however, were such excellent engineers that miles upon miles of their roadway still remain comparatively intact, and where sections are missing it is at least as much due to the ravages of man as the ravages of time.[7]

There is no need to emphasize that Herodotus was not the first man in the Mediterranean lands to have some geologic knowledge. His ability to think regionally must have rested on centuries of preparation.

Before leaving the topic of the geologic knowledge of the ancients, we should seek something of the understanding held in the Orient. Apparently isolation in time and place is not important. The same basic principles of geology emerge whenever a thinker examines the field data with an open mind. Here are some Chinese writings of the neo-Confucian period. The first is by the famous philosopher Chu Hsi (1130-1200 A.D.).

I have seen on high mountains conchs and oyster shells, often embedded in the rocks. These rocks in ancient times were earth or mud, and the conchs and oysters lived in water. Subsequently everything that was at the bottom came to be at the top, and what was originally soft became solid and hard. *One should meditate deeply on such matters, for these facts can be verified* [italics mine].[9]

How similar this statement is to the views we have developed of the sedimentary layers in the Grand Canyon! The four critical geologic aspects are all here, and the least astonishment conclusion is that uplift occurred. Chu Hsi also understood the wasness of the is, for the seashells were correctly assigned to a former time in which these were parts of living animals. Chu Hsi's recognition that the facts can be verified has deep philosophic implications.

Science is a nonpersonal endeavor. The data are the same for everyone. How different this is from many of the more subjective disciplines, in which one man's opinion is not only as good as another's but as an opinion cannot be verified or necessarily duplicated.

Several other Chinese scholars correctly identified fossils and made bold genetic interpretations about the former state of nature in specific localities. Shen Kua in the eleventh century made this observation about the origin of the Asiatic continent.

> When I went to Hopei on official duties I saw that in the northern cliffs of the Thai-Hang Shan mountain-range, there were belts [strata] containing welk-like animals, oyster shells, and stones like the shells of bird's eggs [fossil echinoids]. So this place, though now a thousand li [about 330 miles] west of the sea, must have once been a shore. Thus what we call the "continent" must have been made of mud and sediment which was once below the water. . . .
>
> Now the Great [the Yellow] River, the Chang Shui, the Hu Tho, the Cho Shui and the Sang Chhein are all muddy silt-bearing rivers. In the west of Shensi and Shansi the waters run through gorges as deep as a hundred feet. Naturally mud and silt will be carried eastward by these streams year after year, and in this way the substance of the whole continent must have been laid down. *These principles must certainly be true* [italics mine].[9]

Surely this passage ranks with some of the best of Herodotus. Shen Kua's use of "must" calls on the least astonishment principle after he has set down the list of supporting facts. Paleoecologists may be particularly interested in his reference to a "shore" in relation to the shallow water fossil forms. How striking is his naturalistic realization that continents are formed by an orderly process, and how calm he is about change of place through regional uplift!

Shen Kua wrote another account in which he uses plant fossils and what he knows of similar plants in his own environment as a means of determining the paleoclimate of the remote past. His reasoning is quite similar to that we used in Chapter 4 as we reconstructed a past climate in California on the basis of the redwood rings.

> In recent years [circa 1180] there was a landslide on the bank of a large river in Yung-Ning Kuan near Yenchow. The bank collapsed, opening a space of several dozens of feet, and under the ground a

forest of bamboo shoots was thus revealed. It contained several hundred bamboos with their roots and trunks all complete, and all turned to stone. A high official happened to pass by, and took away several, saying that he would present them to the Emperor. Now bamboos do not grow in Yenchow. These were several dozens of feet below the present surface of the ground, and we do not know in what dynasty they could possibly have grown. Perhaps in very ancient times the climate was different so the place was low, damp, gloomy and suitable for bamboos. . . .[9]

If this has been a successful chapter, the sharp distinction between science and the rest of intellectual thought will begin to blur. In the next chapter we may blur the distinction even further by examining the use of geology in literary imagery. Our purpose is less to humanize science than it is to tie together all of the many ways in which men see and try to comprehend the world.

ANNOTATED REFERENCES

1. Komroff, Manuel (ed.). 1928. The history of Herodotus (translated by George Rawlinson). Tudor Publishing Co., New York. 544 pp. (Herodotus' account of the geology of Egypt and adjacent areas of the Mediterranean is taken from p. 82 ff. Used by permission of Tudor Publishing Co.)

2. Gillespie, Charles Coulton. 1959. Genesis and geology. Harper Torchbooks, The Cloister Library, Harper & Row, Publishers, New York. 306 pp.

3. Biswas, Asi K. 1966. The Nile: its origin and rise. Water and sewage works **113**:282-292. (The information about Nilometers appears on p. 291.)

4. Dunbar, Carl O. 1962. Historical geology (ed. 2). John Wiley & Sons, Inc., New York, 500 pp. (A number of interesting estimates of the age of the earth appear on p. 18 ff.)

5. Aubin, Robert Arnold. 1936. Topographical poetry in XVIII century England. Modern Language Association of America, New York. 419 pp. (Aubin quotes Zerbino's poem on p. 241. Used by permission of the Modern Language Association of America.)

6. Harrington, John W. 1967. The first, first principles of geology. Am. J. Sci. **265**:449-461. (Many of the concepts in this chapter are drawn from this article.)

7. Glueck, Nelson. 1940. On the other side of the Jordan. American Schools of Oriental Research, Cambridge, Mass. 208 pp. (Glueck's discussion of ancient roads [pp. 11-16] and the accompanying illustrations dispel the concept of superhighways constructed with cut-balancing fill in the sixth century B.C. The roads are impressive and have lasted for millenia, but they are not flat. The mountains are rugged and the roads pass over them in a typical winding manner. Used by permission of the American Schools of Oriental Research.)

8. Glueck, Nelson. 1959. Rivers in the desert. Jewish Publication Society of America, New York. 302 pp.

9. Needham, Joseph (with the research assistance of Wang Ling). 1959. Science and civilization in China. Vol. 3. Mathematics and the sciences of the heavens and earth. Cambridge University Press, London. 847 pp. (The quotation from Chu Hsi is found on p. 538 ff; that from Shen Kua is found on pp. 604 ff. and 614 ff. Used by permission of Cambridge University Press.)

Chapter 8

I'm headed for a land that's far away,
   Beside the crystal fountain;
So come with me and you will see,
   The Big Rock Candy Mountain!

Big Rock Candy Mountain, Mac McClintock

The geology in literary imagery

In 1599 Shakespeare's play *King Henry V* was first presented on the cramped quarters of the then new Globe Theater in London. The siege of Harfleur and the Battle of Agincourt could not be shown with the realism of a Hollywood opus. Shakespeare did not have the 150-ampere arc lights used in modern theaters or the advantages of backgrounds filmed on location. The device Shakespeare used to present the grandeur of hordes of men in battle was to transfer the responsibility for visualization from the staging to the imagination of each individual member of the audience. In the Prologue to the play the Chorus enters and dismisses the Hollywood approach with these lines:

O for a Muse of fire, that would ascend
The brightest heaven of invention!
A kingdom for a stage, princes to act,
And monarchs to behold the swelling scene!
Then should the warlike Harry, like himself,
Assume the port of Mars; and at his heels,
Leash'd in like hounds, should famine, sword and fire
Crouch for employment.

111

Then, with a fine turn of language, Shakespeare apologizes for his limitations:

> But pardon, gentles all,
> The flat unraised spirit that hath dar'd
> On this unworthy scaffold to bring forth
> So great an object: can this cockpit hold
> The vasty fields of France? or may we cram
> Within this wooden O the very casques
> that did affright the air at Agincourt?
> O, pardon! since a crooked figure may
> Attest in little place a million . . .[1]

Finally, he skillfully passes the burden of viewing to his audience and catches them up in the spirit of a shared adventure.

> And let us, ciphers to this great acompt,
> On your imaginary forces work.
> Suppose within the girdle of these walls
> Are now confin'd two mighty monarchies,
> Whose high upreared and abutting fronts
> The perilous narrow ocean parts asunder:
> Piece out our imperfections with your thoughts:
> Into a thousand parts divide one man,
> And make imaginary puissance;
> Think, when we talk of horses, that you see them
> Printing their proud hoofs i' the receiving earth;
> For 'tis your thoughts that now must deck our kings,
> Carry them here and there; jumping o'er times,
> Turning the accomplishments of many years
> Into an hour-glass: for the which supply,
> Admit me Chorus to this history;
> Who, prologue-like, your humble patience pray,
> Gently to hear, kindly to judge our play.[1]

It is now time for us "to see a world" through a similar vicarious field experience. We cannot physically meet together and visit the localities. Neither can we use the muse of fire, as Professor Grope calls his 1,000-watt slide projector. Our device will also require us to call on our imaginations to visit places and view pure geologic phenomena through the eyes of poets.

Poetry is particularly adapted to the use of geologic imagery. The eternal cadence of geologic change has attracted many observers through the millenia. Great poets have an ability to record their perceptions in forms as clear and clean as the tooth of a hound. Their nature-linked images frequently contain simple timeless statements of the appearance of things. If his image is accurate, it is not even necessary for the poet to know all about the geologic features he describes. We are sufficiently well trained in scientific analysis to begin to draw our own least astonishment conclusions from any raw data that reflect the true state of nature.

English professors may decry this method. They seem to evolve from a strain of students who shun the science buildings as they would a plague. The world of science is lost to them, and sadly, they to it. Oddly enough this is not true of many of the creative writers for whom professors of literature hold genuine respect. Such a man was Homer as he describes the might of an autumn storm.

There are days in autumn when the whole countryside lies darkened and oppressed under a stormy sky and Zeus sends down torrential rain as a punishment to men. His anger is roused because regardless of the jealous eye of Heaven they have misused their powers, delivered crooked judgements in a public session and driven justice out. In consequence their streams all run in spate, hillsides are scarred by torrents, and the rivers, wrecking the farmlands in their way, rush down headlong from the mountains with a great roar into the turbid sea. Such was the din that went up from the Trojan horse as they fled.[2]

The geology of the sea is also the subject of poetry, for the surf-cadence is as regular as a heartbeat. There is also an interesting bit of reciprocity here, for the modification of shorelines is due to the never-ending pounding of the water every few seconds, not just for days or centuries but for eternity. Three states of matter—liquid, solid, and gas—meet at the shorelines. The gas, or air, is energized by the sun's heat into a convective flow called wind. The energy is translated into the motion of the water and then is absorbed by the work of the destruction of land. The following excerpts and poems (from *Break, Break, Break* by Alfred, Lord Tennyson, a diary entry for October 19, 1958, by Dag Hammarskjøld, an untitled aphorism by Kahlil Gibran, and *Dover Beach* by Matthew Arnold) capture aspects of these processes.

Break, break, break,
On thy cold gray stones, Oh Sea!
And I would that my tongue could utter
The thoughts that arise in me.[3]

Wall of power
In assault,
Wave of light
In the pause,
Then broken,
Receding
From the lip
Of white sand,
Foam and froth.[4]

I am forever walking upon these shores,
Betwixt the sand and the foam.
The high tide will erase my foot-prints,
And the wind will blow away the foam.
But the sea and shore will remain
Forever.[5]

The sea is calm to-night.
The tide is full, the moon lies fair
Upon the straits;—on the French coast the light
Gleams and is gone; the cliffs of England stand,
Glimmering and vast, out in the tranquil bay.
Come to the window, sweet is the night air!
Only, from the long line of spray
Where the sea meets the moon-blanch'd land,
Listen! you hear the grating roar
Of pebbles which the waves draw back, and fling,
At their return, up the high strand,
Begin, and cease, and then again begin,
With tremulous cadence slow, and bring
The eternal note of sadness in.

Sophocles long ago
Heard it on the Aegaean, and it brought
Into his mind the turbid ebb and flow
Of human misery; we
Find also in the sound a thought,
Hearing it by this distant northern sea.[6]

Poets have also used the eternal battle between the land and
the sea to emphasize the effects of change through time. One excellent illustration from Robert Frost's *Once by the Pacific* is paral-
leled in the tourist's view of the California coast (Fig. 15).

The shattered water made a misty din.
Great waves looked over others coming in,
And thought of doing something to the shore
That water never did to land before.
The clouds were low and hairy in the skies,
Like locks blown forward in the gleam of eyes.
You could not tell, and yet it looked as if
The shore was lucky in being backed by cliff,
The cliff in being backed by continent;
It looked as if a night of dark intent
Was coming, and not only a night, an age.
Someone had better be prepared for rage.
There would be more than ocean-water broken
Before God's last *Put out the Light* was spoken.[7]

Frost's illustration is of particular interest because the role of the atmosphere is included in the total picture. The tourist on the cloud-shrouded California coast is immediately totally immersed in all three states of matter. Frost simply saw them all and used them all.

The work of the wind is also featured in Frost's poem *Sand Dunes*. A beach is a localized desert environment produced through

Fig. 15.
Once by the Pacific.

the killing action of salt spray. Without the protective cover of vegetation to hold the soil, the free grains are blown in the wind almost as readily as the ever-free water.

Sea waves are green and wet,
But up from where they die,
Rise others vaster yet,
And those are brown and dry.

They are the sea made land
To come at the fisher town,
And bury in solid sand
The men she could not drown.

She may know cove and cape,
But she does not know mankind
If by any change of shape,
She hopes to cut off mind.

Men left her a ship to sink:
They can leave her a hut as well;
And be but more free to think
For the one more cast off shell.[8]

These themes appear over and over in poetry. It may not be too surprising to see them used by modern writers, for most of them could have been exposed to organized science in one way or another. It is surprising to find that the same themes predate the development of technical geology by at least 200 years. Imagery must be grasped by the audience of folk intellectuals in any period or the writer will fail to communicate his meaning.

The next two selections by Shakespeare are startling. We see him as a conscious observer of geologic change without too much surprise, for genius like Shakespeare's is characterized by a nearly unlimited range of interest. What is surprising is the fact that his imagery was appreciated by his contemporaries, even in the theater pit. Some of the basic truths of geology were known before its full birth as a science. Both of these selections could be reorganized into the prosaic Facts ⟶ Imply via least astonishment ⟶ Conclusion format. Perhaps the most astounding aspect of both selections is that Shakespeare was conscious of the constructive work of the sea in building new land as well as its more familiar destructive role. Both selections, from *Sonnet 64* and *King Henry IV, Part II*, express a straightforward uniformitarianism fully worthy of Hutton.

When I have seen the hungry ocean gain
Advantage on the kingdom of the shore,
And the firm soil win of the wat'ry main,
Increasing store with loss, and loss with store;
When I have seen such interchange of state,
Or state itself confounded to decay;
Ruin hath taught me thus to ruminate—
That time will come and take my love away.[9]

O God! that one might read the book of fate,
And see the revolution of the times
Make mountains level, and the continent,
Weary of solid firmness,—melt itself
Into the sea! and, other times to see
The beachy girdle of the ocean
Too wide for Neptune's hips . . .[10]

There is an interesting parallel between this selection from *King Henry IV, Part II,* and that from Homer's *Iliad;* both poets recognized the erosion of mountains and the consequent deposition in the sea.

Before we leave the sea poems there are two of a different nature that should be included among our examples of lyrically expressed geology. The first, from Shakespeare's *Rape of Lucrece,* might have been posed as a scientific question. Shakespeare the observer has seen an anomaly. How can it be?

The petty streams that pay a daily debt
To their salt sovereign, with their fresh falls' haste,
Add to his flow, but alter not his taste.[11]

Another poem by Gilbran also poses a question. The subject once again appears to be the chemistry of sea water. Actually, the poet is concerned with the story of organic evolution and its sequence of changes through time. He wrote in a time (circa 1925) when evolution had been accepted in scientific circles as reality. Vertebrates became terrestrial rather than simply marine animals in Devonian time although all their body juices were still tuned to the earlier saline marine environment. For the last 360 million years or more we have manufactured our own sea water. Gibran, in *Sand and Foam,* says it so much better.

There must be something strangely
sacred in salt. It is in our tears and in the sea.[12]

Gibran's words live because he expresses so many universally shared emotions so much better than anyone else. His view of time is fully as profound as that of Herodotus and Shen Kua.

We measure time according to the
movement of countless suns; and they
measure time by little machines in their
little pockets.

Now tell me, how could we ever meet
at the same place and the same time?[13]

Professor Grope has often felt this way as he lectures on Cambrian trilobites to a class of freshmen, especially on Friday just before 11:00 A.M. when the weekend hangs suspended on a line of minutes that must pass before the bell sounds.

The next poem, *Description of Bath*, is especially interesting because, although it was written in 1734, it anticipates a knowledge of rocks and time that would have been remarkable a full century later. The author is the lively little milliner, Mrs. Mary Chandler. Please explain how a hatmaker knew this.

The shatter'd *Rocks* and *Strata* seem to say,
"Nature is *old*, and tends to her *Decay*":
Yet, *lovely in Decay*, and green in *Age*,
Her Beauty lasts her to her *latest Stage.*[14]

To someone who has never seen a glacier, ice en masse may seem a strange topic for a poem. On the contrary, it is easier to conceive of glaciers in poetic terms than in the frigid words of textbook prose. *Mont Blanc*, by Shelley, presents his unique perception of a living glacier.

Mont Blanc appears—still, snowy, and serene—
Its subject mountains their unearthly forms
Pile around it, ice and rock; broad vales between
Of frozen floods, unfathomable deeps,
Blue as the overhanging heaven, that spread
And wind among the accumulated steeps;
A desert peopled by the storms alone,
Save when the eagle brings some hunter's bone,
And the wolf tracks her there—how hideously
Its shapes are heaped around! rude, bare and high,

Ghastly, and scarred, and riven—Is this the scene
Where the old Earthquake-daemon taught her young
Ruin? Were these their toys?

Power dwells apart in its tranquillity—
Remote, serene, and inaccessible:
And *this,* the naked countenance of earth,
On which I gaze, even these primeval mountains
Teach the averting mind. The glaciers creep
Like snakes that watch their prey, from their far fountains
Slow rolling on; there, many a precipice,
Frost and the Sun in scorn of mortal power
Have piled: dome, pyramid and pinnacle,
A city of death, distinct with many a tower
And wall impregnable of beaming ice.[15]

Coleridge also wrote a poem that is a magnificent description
of a glacier—*Hymn Before Sunrise in the Vale of Chamouni.* The first
portion of this poem deals with the mad rush of meltwater as it
issues from the glacier's front. Professor Grope is reminded of a
day he spent in British Columbia on the Kicking Horse River,
which plunges white with glacial milk down the western slope of
the Canadian Rockies. Frontiersmen often chose names that are
equally poetic, for example, Mule Shoe and No Trees, both in
Texas.

From dark and icy caverns called you forth,
Down those precipitous, black, jagged rocks,
Forever shattered, and the same for ever?
Who gave you your invulnerable life,
Your strength, your speed, your fury and your joy,
Unceasing thunder and eternal foam?[16]

In the same poem Coleridge expresses an idea that has occurred
to many practicing geologists who have thought about the unusual
climatic changes that have happened during the last two million
years.

And who commanded (and the silence came),
Here let the billows stiffen, and have rest?
Ye Ice-falls! Ye that from the mountain's brow
Adown enormous ravines slope amain—
Torrents, methinks, that heard a mighty voice,

And stopped at once amid their maddest plunge!
Motionless torrents! silent cataracts!
Who made you glorious as the Gates of Heaven. . .[16]

These two poems are of unusual historical interest, for Shelley (1791-1822) and Coleridge (1772-1834) were both dead before geologists paid much attention to glacial action. Louis Agassiz (1807-1873) was the best-known early geologist to discover the meanings of glacial flow. In about 1837 he realized that the erosional and depositional features found in the Swiss valleys below the ice fronts signified a change in climate from some colder time when there was more ice than in the present climate. He also saw that similar features found across the plains of northern Europe and in the northern portions of the British Isles were formed by ice action.

Agassiz's theory of glaciation began to receive serious attention after 1840. Geologists had difficulty in comprehending that a mass of ice larger than the sheet that covers Greenland lay over the whole of northern Europe just a few tens of thousands of years earlier. Agassiz and other geologists finally proved that continental glaciation had occurred in both Europe and North America. One of the most dramatic proofs of ice flow and erosion is found in the scratched, scraped, and scoured bedrock floor. Flowing glacial ice that contains rocks frozen into the mass acts as a rasp and is capable of scratching an entire countryside as no other agent can. Agassiz experimented with measurements of glacial flow as a youth when he drove stakes in a line across a glacier and watched progressive deformation. Shelley also knew about flow, for he said, "The glaciers creep/Like snakes that watch their prey, from their far fountains/Slow rolling on . . ."

Robert Frost wrote of glaciers too, but in a less direct way. His New England countryside had been glaciated by the American ice sheet that overrode the mountains and left a blanket of boulders across the land. For centuries Yankee farmers have made their classic stone walls from this boulder debris that was abandoned as the ice melted away. There are a number of lines of pure geologic imagery in his poem *Mending Wall*. Geologists call this weathering action "frost heaving" (a fact, not a pun).

Something there is that doesn't love a wall,
That sends the frozen ground-swell under it,
And spills the upper boulders in the sun;
And makes gaps even two can pass abreast.[17]

These glacial boulders are described a few lines farther down.

We keep the wall between us as we go.
To each the boulders that have fallen to each.
And some are loaves and some so nearly balls
We have to use a spell to make them balance:
"Stay where you are until our backs are turned!"[17]

There are an almost unlimited number of river poems containing remarkable geologic insights. A typical example, the spectacular action of a flash flood, is given in another of Frost's poems, *One Step Backward Taken*. Professor Grope heard Frost describe how he happened to write it. During a train ride across the western desert, a torrential wash from a thunderstorm threatened to cut away the railroad embankment. Peering out of the train window, Frost saw what was happening. The train pulled back until it was safe to go on, and the episode helped the poet to solve a personal crisis.

Not only sands and gravels
Were once more on their travels,
But gulping muddy gallons
Great boulders off their balance
Bumped heads together dully
And started down the gully.

Whole capes caked off in slices
I felt my standpoint shaken
In the universal crisis.
But with one step backward taken
I saved myself from going.
A world torn loose went by me.
Then the rain stopped and the blowing
And the sun came out to dry me.[18]

There is a link between this poem and Isaiah 40:4 that we might not suspect unless we knew that the incident had occurred in the Basin and Range Province. The lines "Not only sands and gravels/ Were once more on their travels" tell of Frost's awareness that erosion, transportation, and deposition by streams is a story of step-by-step movement as each high-water period moves things a little further, until to quote Luke (3:5), "Every valley shall be filled/And every mountain and hill shall be brought low."

*   *   *

Poetry must be read with so much concentration that it is hard
to assimilate in bulk; therefore before tackling the next classic it
may be well to present a short prose selection as a change of pace.
The geology in both cases is similar, for it deals with the long
profile of a major river and the changes that occur from headwaters
to discharge.

Much of a stream's behavior is only revealed by studying its entire
length—particularly its changes in gradient, stream pattern, discharge
and load—from the headwaters to the mouth. . . . The long profile [also
called longitudinal profile] of a stream is a graphic outline of the
stream's gradient over long segments of its course. . . . The long profile of the Arkansas River is representative of many streams, although
each one differs in profile from every other, not only because of
differences in discharge, load and other factors mentioned, but also
because of more subtle differences related to bedrock structure and
geologic history of the area traversed. The Arkansas River rises in the
southern Rocky Mountains and flows across the Great Plains, joining
the Mississippi about 440 miles from the Gulf of Mexico. Above Canon
City, Colorado, the gradient is steep and irregular; the river plunges
swiftly through deep mountain canyons, including the awe-inspiring
chasm of the Royal Gorge cut deeply into the resistant granites and
gneisses of the Front Range. By contrast the gradient of the lower 186
miles of the river is very regular, nowhere greater than about 12 inches
a mile: here the river winds in serpentine bends on a gently sloping
plain of sand and silt—material almost exactly like the load now being
carried by the stream. Between these extremes the gradient of the
middle course of the river varies regularly and gradually, so that the
entire profile approximates a smooth concave-upward curve that
steepens notably near the river's head. The entire profile roughly resembles a segment of a hyperbola, but in detail it is far from a simple
mathematical curve.[19]

Sidney Lanier's poem *Song of the Chattahoochee* opens with a description of the headwaters region of one of the oldest rivers in
America. The river flows southeast out of the Blue Ridge Mountains on the western side of the continental divide in Georgia. Then,
on reaching the elevated plain of the Dalhonega Plateau in Hall
County, it turns almost due south. The eastern continental divide
passes along Peachtree Street in downtown Atlanta. Gutter drainage on the eastern side flows into the Atlantic Ocean through a
rather young system of rivers, for the ocean itself is rather young.
Drainage on the west side flows into the much more ancient Chat-

tahoochee. In the Atlanta area the Chattahoochee is still flowing across crystalline rocks comparable to those of the inner gorge of the Grand Canyon. At Columbus, Georgia, the river reaches the coastal plain, an area of sediments deposited within the last hundred million years. From this point, the river is a younger extension of the old pioneer. As the coastal plain was lifted to its present position above sea level, the river continued to flow southward to reach the gulf. The Flint River joins the Chattahoochee at the Florida-Georgia state line. Below this point the system is called the Apalachicola River.

Lanier paints a picture in changing rhythm and pace so that the reader can tell exactly where he is on the long profile. The textbook description ends with the line "but in detail it is far from a simple mathematical curve." Hidden in those words is the recognition that the profile of the headwaters is full of falls and rapids with a stair-step descent. This is how the poet begins, with the spritely verbs "hurry," "run," "leap," "split," "flee"; then he slows with the dragging out of rhythm and meaning in the line, "With a lover's pain to attain the plain. . . ."

Out of the hills of Habersham,
Down the valleys of Hall,
I hurry amain to reach the plain,
Run the rapid and leap the fall,
Split at the rock and together again,
Accept my bed, or narrow or wide,
And flee from folly on every side
With a lover's pain to attain the plain
Far from the hills of Habersham,
Far from the valleys of Hall.

And oft in the hills of Habersham,
And oft in the valleys of Hall.
The white quartz shone, and the smooth brook-stone
Did bar me passage with friendly brawl,
And many a luminous jewel lone
—Crystals clear or a-cloud with mist,
Ruby, garnet, and amethyst—
Made lures with the lights of the streaming stone
In the clefts of the hills of Habersham,
In the beds of the valleys of Hall.

But oh, not the hills of Habersham,
And oh, not the valleys of Hall

Avail: I am fain for to water the plain.
Downward the voices of Duty call—
Downward, to toil and be mixed with the main.
The dry fields burn, and the mills are to turn,
And a myriad flowers mortally yearn,
And the lordly main from beyond the plain
Calls o'er the hills of Habersham,
Calls through the valleys of Hall.[20]

From a geologic standpoint, the minerals named reveal the high pressure and temperature environment under which the rocks of the headwaters area were formed. Many miles of overburden had to be stripped away by the progressive erosion of this river system before the veins containing milky quartz and amethyst could be exposed. Garnets and rubies record an even higher pressure environment. This is a duty poem in the finest Southern tradition. Two other verses picture the botanic beauty of Georgia hill country as hauntingly as the memory of any Georgia boy could duplicate.

Engineering geology should not be left out, and our example was written by Jonas Moore, who discussed the geology of France and England with Samuel Pepys, as we noted in Chapters 2 and 4. You may remember that Moore was Surveyor General of Ordnance under Charles II and was responsible for supervising the work of draining the Fens, a large swampy area in East Anglia. This project, begun by the Romans, is still in progress. It is made difficult by the collapse of glacial peat bogs as they are de-watered. In places subsidence of this land has lowered it below the levels of the drainage streams and canals. In the last part of the seventeenth century the modern drainage project was begun by Jonas Moore with the high hopes expressed in these lines from *The History or Narrative of the Great Level of the Fens, Called Bedford Level:*

I sing Floods muzled, and the Ocean tam'd
Luxurious Rivers govern'd, and reclaim'd
Waters with Banks confin'd, as in a Gaol,
Till kinder Sluces let them go on Bail;
Streams curb'd with Dammes like Bridles, taught t' obey,
And run as strait, as if they saw their way.[21]

One last thought is necessary. There are no real rocks in the pages of a book. There are no streams, no mountains, no glaciers or volcanos, only words expressing the opinions of some human being. These are just abstractions of the real world. In the choice between the two, read the course Walt Whitman charts in *When I Heard the Learn'd Astronomer,* and follow the poet into the night:

When I heard the learn'd astronomer,
When the proofs, the figures, were ranged in columns before me,
When I was shown the charts and diagrams, to add,
    divide, and measure them,
When I sitting heard the astronomer where he lectured
    with much applause in the lecture-room
How soon unaccountable I became tired and sick,
Till rising and gliding out I wander'd off by myself,
In the mystical moist night-air, and from time to time,
Look'd up in perfect silence at the stars.[22]

ANNOTATED REFERENCES

1. Shakespeare, William. King Henry V. In The complete works of William Shakespeare. No date. Walter J. Black, Publisher, New York. 1312 pp. (The Prologue to the play is found on p. 531.)

2. Rieu. E. V. (translator). 1950. The Iliad (by Homer). Penguin Books, Inc., Baltimore. 469 pp. (This selection is taken from Book XVI, "Patroclus fights and dies," p. 303. Used by permission of Penguin Books, Inc.)

3. Tennyson, Alfred, Lord. 1887. Poetical works. Mims & Knight, Troy, N. Y. 742 pp. (This poem is found on p. 122.)

4. Hammarskiøld. Dag. 1966. Markings (translated from the Swedish by Lief Sjöberg and W. H. Auden). Alfred A. Knopf, Inc., New York. 222 pp. (This poem is found on p. 172. Used by permission of Alfred A. Knopf, Inc.)

5. Gibran, Kahlil. 1926. Sand and foam. Alfred A. Knopf, Inc., New York. 85 pp. (This is the first aphorism in the book, and it is found on p. 1. Used by permission of Alfred A. Knopf, Inc.)

6. Arnold, Matthew. 1902. Poems of Matthew Arnold. P. F. Collier & Sons, New York. 466 pp. (This poem is found on p. 212.)

7. Frost, Robert. Once by the Pacific. In Latham, Edward Connery (ed.). 1969. The poetry of Robert Frost. Holt, Rinehart & Winston, Inc., New York. 607 pp. (Copyright 1928, 1930, 1939, 1947, and 1969 by Holt, Rinehart & Winston, Inc. Copyright 1956 and 1958 by Robert Frost. Copyright 1967 by Lesley Frost Ballantine. Used by permission of Holt, Rinehart & Winston, Inc., and the Estate of Robert Frost.)

8. Frost, Robert. Sand dunes. In Latham, Edward Connery (ed.). 1969. The poetry of Robert Frost. Holt, Rinehart & Winston, Inc., New York. 607 pp. (Copyright 1928, 1930, 1939, 1947, and 1969 by Holt, Rinehart & Winston, Inc. Copyright 1956 and 1958 by Robert Frost. Copyright 1967 by Lesley Frost Ballantine. Used by permission of Holt, Rinehart & Winston, Inc., and the Estate of Robert Frost.)

9. Shakespeare, William. Sonnet 64. In The complete works of William Shakespeare. No date. Walter J. Black, Publisher, New York. 1312 pp. (This poem is found on p. 1256.)

10. Shakespeare, William. King Henry IV, Part II. In The complete works of William Shakespeare. No date. Walter J. Black, Publisher, New York. 1312 pp. (This excerpt from Act III, scene i, lines 45-51, is found on p. 511.)

11. Shakespeare, William. The rape of Lucrece. In The complete works of William Shakespeare. No date. Walter J. Black, Publisher, New York. 1312 pp. (These lines, 649-651, are found on p. 1231. Lines 589-592 on the same page contain another interesting geologic reference.)

    All which together, like a troubled ocean,
    Beat at thy rocky and wreck-threat'ning heart;
    To soften it with their continual motion;
    For stones dissolv'd to water do convert.

12. Gibran, Kahlil. 1926. Sand and foam. Alfred A. Knopf, Inc., New York. 85 pp. (This aphorism is found on p. 79. Used by permission of Alfred A. Knopf, Inc.)

13. Gibran, Kahlil. 1926. Sand and foam. Alfred A. Knopf, Inc., New York. 85 pp. (This passage is found on p. 7. Used by permission of Alfred A. Knopf, Inc.)

14. Aubin, Robert Arnold. 1936. Topographical poetry in XVIII century England. Modern Language Association of America, New York. 419 pp. (Aubin quotes Mary Chandler's poem on p. 164. Used by permission of the Modern Language Association of America.)

15. Shelley, Percy Bysshe. Mont Blanc. In Page, Curtis Hidden (ed.). 1910. British poets of the nineteenth century (ed. 4). B. H. Sanborn & Co., New York. 935 pp. (This poem is found on pp. 288-291.)

16. Coleridge, Samuel Taylor. Hymn before sunrise in the vale of Chamouni. In Page, Curtis Hidden (ed.). p. 110. British poets of the nineteenth century (ed. 4). B. H. Sanborn & Co., New York, 935 pp. (This poem is found on pp. 96-98. Both Shelley and Coleridge were inspired by visits to the Chamouni Valley on the northwest side of Mont Blanc.)

17. Frost, Robert. Mending wall. In Latham, Edward Connery (ed.). 1969. The poetry of Robert Frost. Holt, Rinehart & Winston, Inc., New York. 607 pp. (Copyright 1928, 1930, 1939, 1947, and 1969 by Holt, Rinehart & Winston, Inc. Copyright 1956 and 1958 by Robert Frost. Copyright 1967 by Lesley Frost Ballantine. Used by permission of Holt, Rinehart & Winston, Inc., and the Estate of Robert Frost.)

18. Frost, Robert. One step backward taken. In Latham, Edward Connery (ed.). The poetry of Robert Frost. Holt, Rinehart & Winston, Inc., New York. 607 pp. (Copyright 1928, 1930, 1939, 1947, and 1969 by Holt, Rinehart & Winston, Inc. Copyright 1956 and 1958 by Robert Frost. Copyright 1967 by Lesley Frost Ballantine. Used by permission of Holt, Rinehart & Winston, Inc., and the Estate of Robert Frost.)

19. Gilluly, James, Waters, Aaron C., and Woodford, A. O. 1968. Principles of geology (ed. 3). W. H. Freeman & Co., San Francisco. 678 pp. (This prose selection is found on pp. 221-222. It is good scientific writing, but you cannot hear the river as you can in Lanier's poem. Used by permission of W. H. Freeman & Co.)

20. Lanier, Sidney. Song of the Chattahoochee. In Lanier, Mary D. (ed.). 1892. Poems of Sidney Lanier. Charles Scribner's Sons, New York. 260 pp. (This poem is found on page. 24.)

21. Aubin, Robert Arnold. 1936. Topographical poetry in XVIII century England. Modern Language Association of America, New York. 419 pp. (Jonas Moore is a poet, too! This glimpse of an unusual man is found on p. 194. Used by permission of the Modern Language Association of America.)

22. Whitman, Walt. 1900. Leaves of grass. David McKay Co., Inc., New York. 526 pp. (This poem is found on p. 341.)

Chapter 9

Isn't it astonishing that all these secrets
have been preserved for so many years
just so that we could discover them!

Orville Wright[1]

Developing your geologic creativity and insight

The purpose of this book is to help each reader "to see a world" once hidden from view by encouraging the development of creativity and insight rather than by listing information. No teacher can claim the key that will transform an osteocephalate into a da Vinci incarnate. Yet it may be possible to encourage most people to think creatively. What is needed is a technique by which the magical joy of creative thinking may be recognized instantly for what it is, thus preventing our minds from remaining entrenched in old patterns.

Joseph Kestin has found that part of the process of creative thinking is analogous to perceiving the punchline of a joke.[2] Since the ability to "get" a joke is so common, this linkage may prove to be a very important tool in developing creative insight in nearly everyone. Here is how it works.

An Englishman once came to New York and was shown the sights by a proud American friend. Inevitably they ended up on the street outside the Empire State Building. As they looked up, the American said, "There it is, 102 stories high and completely fireproof!" The Englishman gazed in wonder and then murmured softly, almost to himself, "What a pity."

Now who among us would have ever thought of that? Yet who among us ever again will look at a skyscraper and not chuckle in fiendish delight at thought of such a fire. Responding to a joke is a very personal creative adventure; no one can grasp the point for someone else. The highly personal nature of creativity may be demonstrated quite readily. How many times have you heard a joke that put others in stitches and yet did not seem at all funny to you? Then, some time later, your mind made the creative leap and you, too, were convulsed. Creativity is hard to control. The leap may be made at a most inappropriate time, where it frequently takes precedence over anything else that is going on. This common experience proves the point quite well.

As Kuhn (see Chapter 2) has shown, the creative process in science follows some individual's ability to see nature in a different way. The basic data of the field are known to all the practitioners, but they view it in much the same way that one listens to the preamble of a joke. The mental processes involved in getting a joke and making a scientific breakthrough are nearly identical. Once some observer has been able to look at nature differently, each practitioner must still duplicate that thinking for himself before he is also able to use the breakthrough.

In this final chapter we will look at some geologic data within the rigid framework of familiar factual information. All the data involve the continent of North America and are keyed to Fig. 16. The presentation follows the same format used for telling a joke. Facts 1 through N are given to set the stage. Then the next paragraphs, which are set off under the heading "Therefore, via least astonishment," contain conclusions through which the facts may be seen in a new way. If this method of developing creative insight is successful, each reader should enjoy a moment of glee as he grasps the conclusions. This sense of fun is an earmark of the philosophic approach. This is the technique by which the magical joy of creative thinking may be recognized immediately. Here we go!

Example, facts 1 to N:

The entire set of examples deals with rivers and lakes. For this purpose Fig. 16 was drawn as an outline map showing all the rivers and major lakes (see foldout at back of book). The map was simplified by purposely omitting other physical features such as mountains. Shorelines are somewhat simplified, omitting minor islands and inlets. Drainage, as might be expected, is essentially radial, as the rivers flow from the interior of the continent as spokes radiate from the axle of a wheel, reaching the coast of every ocean, gulf, and bay. These are the simple facts.

Therefore, via least astonishment:

Since water flows downhill, the inland ends of the river branches are the highest parts of the bottoms of each drainage basin. The divides between river systems, although not shown on the map, may be inferred as the highest parts of the rims of the drainage basins. At first this may not seem to be a meaningful statement. However, consider what the map discloses about the position of the continental divide in Florida and Georgia or, for that matter, in the rest of the country. The crest dividing the eastern drainage into the "young" Atlantic Ocean (180 million years old) from the western drainage into the much older Gulf of Mexico is only about 150 feet above sea level throughout the length of Florida. The continental divide continues inland across the Georgia Piedmont and at one point is occupied by the most important street in downtown Atlanta. Peachtree Street lies along the crest between the Chattahoochee River drainage and the Altamaha River drainage (CH.R., AL.R.). At this point the continental divide has an elevation of about 1,000 feet. We think of continental divides as passing among snowy peaks, and they do, but they must also come to the sea as illustrated in Florida.

Example, facts 1 to N:

The geologic structure of Florida is an anticlinal warp with a crest of the fold paralleling the length of the peninsula. The rocks are chiefly marine sediments that were deposited over the buried crystalline basement at depths of from less than a mile to nearly three miles. Limestones are being deposited today along the western and southern portions of the arch just as they were in the center area in the late Eocene epoch (about 40 million years ago). Florida has a very humid climate, with an average rainfall of between 50 and 60 inches per year. Pure limestone weathers rapidly if there is sufficient water and suitable underground plumbing to allow the water to flow away through the rocks to make room for new and unsaturated solvent. There is an abundance of lakes in the limestone portions of the Florida arch, but easy subsurface drainage to the sea is hidden. The state seems too flat and low for large-scale subsurface drainage. During the Pleistocene epoch that began about 2 million years ago and ended about 20,000 years ago, a great deal of former sea water was locked up on the land in high latitudes as glacial ice. The sea level was lowered by 325 to 470 feet, so that Florida stood as a reasonably high ridge. Fresh rainwater in the cracks of the limestones exerted a pressure of about 180 pounds per square inch at the base of a saturated column extending to what was then sea level.

Therefore, via least astonishment:

The hundreds of lakes of all sizes simply fill holes where limestone was dissolved away, principally during periods of maximum glaciation. The lakes are as much a part of the Pleistocene scenery of Florida as they are a part of the modern world. Florida has few rivers because so much of the water flows off underground through caverns. It reminds us of Coleridge's *Xanadu:*

Where Alph, the sacred river, ran
Through caverns measureless to man,
Down to a sunless sea.

Even reference to the sunless sea is quite apt. Under this hydrostatic head, fresh water could escape into the sea against the pressure of the more dense sea water at depths of well over a mile below sea level. It is truly dark down there.

Example, facts 1 to N:

There are a multitude of lakes in the northern part of the United States and Canada. These are concentrated in the area north of the heavy dashed line in Fig. 16. This line marks the position of the southern limit of glaciation during the Pleistocene epoch. Only a few tens of thousands of years ago a great ice sheet covered the continent north of this line. Comparisons with the present ice sheets over Greenland and Antarctica indicate that the North American sheet was also as much as two miles thick. The amount of ice was dependent on two factors: the yearly gain due to the addition of snow and sleet and the yearly loss due to melting and evaporation. Naturally the area near the Arctic Circle was colder than the areas in the vicinity of the present positions of the Ohio and Missouri Rivers (O.R., MR.R.). Ice is a plastic substance, and large masses of it will flow under their own weight once they become about 70 feet thick. Flowing ice with rocks locked into it like abrasive teeth on a file is a very effective agent of erosion. Imagine the efficiency of such a tool pressed down by the weight of a mile or two of ice and flowing slowly across the bedrock floor!

Therefore, via least astonishment:

Obviously the northern lakes, including those we call the Great Lakes, are filled basins that were scooped from rocks of different degrees of resistance to the abrasion of flowing glaciers. The scale of these excavations almost defies description. Lake Superior (L.S.), with a floor 1,333 feet below the lake-level elevation of 602 feet, is the deepest of the Great Lakes. The floor is actually 731 feet below sea level! Lake Michigan (L.M.) has a surface elevation of 580 feet and a depth of 932 feet. Lake Huron (L.H.) also has a surface elevation of 580 feet but its depth is only 750 feet. This is still considerably below sea level. Lake Erie (L.E.), the shallowest, has a depth of only 210 feet below a lake-level elevation of 572 feet. Lake Ontario (L.O.) discharges into the St. Lawrence River from an elevation of 246 feet. The depth of Lake Ontario, 778 feet, places its bottom well below sea level too. The power of continental glaciation is unmatched as a digger of holes!

Example, facts 1 to N:

These lakes are all interconnected by streams, channels, or straits. Herodotus has taught us to see such features as a class of phenomena rather than as individual elements. Any lake is a temporary geologic feature that may disappear in one of three different ways. In dry climates, lake waters may evaporate. During the process of evaporation, matter dissolved in the water precipitates out, leaving the lake bed as a salt flat. Lakes may become filled with the sediment brought into them by streams, or they may be drained when the dam that contains the mass of their water is eroded.

Therefore, via least astonishment:

The future of the Great Lakes poses an interesting problem. Only Lake Erie is capable of being drained entirely by the erosion of the dam. The others are too deep. Erosion can cut down no further than the permanent base level of the sea, so the deeper lakes are safe from that fate at least. Lake Erie at the present time is being drained at a very rapid rate. The Niagara River connects Lake Erie (elevation 572 feet) with Lake Ontario (elevation 246 feet) and tumbles over Niagara Falls (N.F.) in the process. The waterfall is supported by a cap of resistant limestone that in turn overlies a mechanically weak shale. Erosion below the falls is very active and cuts the shale away behind the water curtain, so that the caprock is unsupported as an overhanging ledge. As the cap breaks off, the waterfall retreats upstream, leaving a narrow gorge in its wake. The Niagara gorge is already about 7 miles long, even though the activity has been going on for no more than 10,000 years. The rate of retreat is on the order of four feet per year. Engineers may be able to slow this rate down considerably, but inevitably the constant retreat of the falls may be expected to drain Lake Erie in the next 30,000 years or so. When this occurs, the Lake Erie basin will be a river valley and the Niagara River will extend as far west as Detroit, Michigan.

Example, facts 1 to N:

One of the most astounding facts derived from Fig. 16 is the almost complete lack of drainage into the Great Lakes. The river drainage is away from the region rather than into it as one might expect.

Therefore, via least astonishment:

Very little sediment is being deposited on the floors of these basins. Therefore it will be quite some time before the deep lakes fill up. Past, present, and future are hard to separate in geology, for they share so many features. When the glaciers melted away, the first water in the lakes was made of melted ice. Unless convective overturn and mixing with modern rains has been complete, there is probably a good supply of the original water still left in the deep holes. No wonder there are such acute pollution problems in these lakes. The water interchange is too limited!

To be strictly honest, we must admit that other factors have interfered with this simple view. When the glaciers first began to melt, the weight of ice on the continent was still sufficient to press down on the surface and depress the Champlain area below sea level. For a short time there was a connection between the Great Lakes Basin and the Atlantic Ocean. Mollusk fossils and even whale bones found as far inland as Michigan mark this period. As the melting of the ice continued into recent times, the load that was holding down the crust disappeared and the land rose to its present position. It was at this time that the water was actually and finally locked into the deep holes.

Example, facts 1 to N:

Nearly half of North America lies north of the line marking the southern limit of the continental ice cap.

Therefore, via least astonishment:

In this entire area there is not a river that is more than a few tens of thousands of years old. These are the youngest rivers in the world, comparable only to similar youthful streams in the other glaciated regions. (Perhaps the controversy over the Loch Ness monster seems less serious when viewed in these terms. Scotland was also glaciated and Loch Ness must have suffered some of the effects.) This is still lake country only because there has been too little time for the erosion of the thousands of hard rock dams. The drainage is simply not well established. Gradually over the next million years or so we may expect the dams to be worn away and simpler stream patterns to emerge.

Example, facts 1 to N:

What of the region south of the area that was covered by the continental ice cap? How ancient are the rivers that lie south of the glacial limit? The Ohio and Missouri Rivers (O.R., MR.R.) have complicated histories because of their positions along the limit of the ice front. For the most part they, too, are young rivers that have taken the place of ancestral streams that were lost when the ice masses built up over the older valleys. The Mississippi River (MS.R.) is the better example, for it may be viewed more clearly on the continental scale. Its present bed, south of the limit of the ice cap, is laid out over a great sheet of outwash debris. As the glaciers melted away, the muddy water flowed off to the south through this channel. A mass of sand, gravel, silt, and clay, several hundred feet thick and in places more than 50 miles wide, was deposited all the way down to the Gulf of Mexico. Naturally a great delta system was formed as the river water slowed to an ocean pace.

Therefore, via least astonishment:

The Mississippi River delta is actually a composite of several older units as well as the section that is growing to the south and east today. During great floods the river tends to flow as a sheet across the natural levees of the delta and into the gulf in many directions. A new channel may be cut during these times, and the whole process of delta building will then shift to that new direction as the older area is abandoned. Not long ago the Mississippi discharged to the southwest, and the abandoned channels of that structure are still visible from the air. Lake Pontchartrain (L.P.) is a relic of a former arm of the gulf that was cut off as a part of the delta built southeastward, closing the southern end of the lake. The Chandeleur Islands (CD.I.) are all that now remain above the sea of another old and abandoned delta unit. Wave action has done its work in a manner beautifully described in Shakespeare's *Sonnet 64*, quoted in Chapter 8.

Example, facts 1 to N:

The Mississippi River delta is large for two reasons. The Mississippi itself is a large river, but this alone does not account for its delta. The river discharges 630,000 cubic feet of water every second; this is hardly comparable to the 4,000,000 to 5,000,000 cubic feet of water discharged by the Amazon; yet the Amazon does not have a protruding delta.

Therefore, via least astonishment:

The currents of the Gulf of Mexico are not capable of distributing the Mississippi sediment as rapidly as it is added. This, combined with the tremendous load of glacial debris that was brought down in a short period of time, has more to do with the delta than the absolute volume of flow. In these terms the deltas of the Rio Grande (RG.R.) and the Apalachicola River (A.R.) are respectable deltas and are comparable to those cited by Herodotus in Greece and Turkey.

Example, facts 1 to N:

One simple truth must be recognized before the concept of the antiquity of rivers has much meaning. An area beneath the sea is not subject to simple stream erosion from the runoff of seasonal rains, but any former sea floor that is raised above sea level must bear streams and exhibit their effects immediately. The date of the last deposition of marine sediment is a key to the age of the rivers flowing over it. The full story of the history of rivers may not be read so simply after a few tens to hundreds of millions of years of stream action. Erosion removes some of the information on which dating may be based. Although the rocks of the midcontinent area may date from the pre-Cambrian period (more than 600 million years ago) through the Paleozoic era (ending some 225 million years ago), the ages of those rivers are not determined solely on the basis of the dates of the youngest marine beds. The detailed story for each area is too laborious to fit into a single chapter of this sort. The approach given here is a little simplified and concentrates on the last hundred million years, with the admission that many of the streams are much older than the minimum dates that might be assigned to them.

There is a dotted line in Fig. 16 extending from the Rio Grande into southern Illinois and back around the eastern states to New

York City. This line marks the present limit of the Eocene marine sediments of the Atlantic and Gulf coastal plains and the Mississippi embayment. The line is not exactly congruent with the shoreline of the depositing Eocene seas about 55 million years ago, but it is within a few tens of miles of the correct position.

Therefore, via least astonishment:

Obviously uplift has occurred, and all the rivers that once emptied into the seas whose borders are roughly approximated by the dotted line have been extended across the new land to the present position of the shore. The Mississippi River has made the longest extension, from the approximate location of Cairo, Illinois, to the delta. All of this portion of the Mississippi River is younger than the part between the dotted (Eocene) line and the line that marks the limit of the glacial ice cap. The older segment may be nearly 275 million years old, possibly more. From the Rio Grande all the way around to New York, the rivers cross this line at approximately right angles. This is natural, for each of the rivers once reached the sea at these points by flowing down a simple regional slope. That rivers like the Red (R.R.) and Brazos (B.R.) of Texas first flowed directly to the sea is shown by the way in which they meet the Eocene shoreline at right angles. Then they change direction and flow down a younger slope to the shoreline of the present Gulf of Mexico, meeting it at right angles also. This behavior is particularly well illustrated by the Red River. It flows in a nearly west-to-east line between Texas and Oklahoma and then bends sharply south at Texarkana to enter the Gulf of Mexico as the Atchafalaya River. The upper courses of the rivers crossing the dotted Eocene Sea level line are all more than 55 million years old, while the lower courses must be younger than that.

The Sabine River (SB.R.) of Texas is an example of a class of rivers that must be young because their entire courses are laid out on very young rocks. The Sabine flows across elevated, subaerially exposed marine sediments that date from the Eocene epoch or later. We are not used to looking at rivers and thinking about their histories, but we can learn to do it!

Example, facts 1 to N:

Rivers extend the areas of their drainage basins by means of headward erosion. At the upper ends of every major river system there are hundreds of small tributaries flowing on steep gradients down a series of falls and rapids. Headward erosion is accomplished by the retreat of these falls as their caprocks break off in the manner of tiny Niagaras. The rapid and destructive erosion of farm fields is a familiar example of the process. Flowing destroys the caprock strength of tough sod, and fingering valleys are then extended headward with each gully-washing rain. The same processes exist on a grand scale between all river systems separated by ridgelike divides. Naturally the overnight effects are not so noticeable as they are on farm fields, but the struggle between rivers is never ending. If one river gains drainage basin territory, the adjacent river must lose it.

Old rivers that have been in competition with one another for hundreds of millions of years have had time to stabilize, so that the rate of further change is quite slow. The activities of these rivers are held in check by constraints downstream. The New River (N.R.) is a part of the headwaters of the Mississippi-Ohio River system that extends well back into the Blue Ridge portion of the Appalachian Mountains near Boone, North Carolina. The New River flows northeastward out of the heart of Watauga County along the line of the structural grain of the mountain rocks before turning to the northwest in Virginia and joining the Kanawah River at Charleston, West Virginia. The Kanawah flows into the Ohio, so North Carolina mountain water, after passing through a number of northern states and finally turning south, eventually reaches the Gulf of Mexico. The upper reaches of the New River are about 300 million years old. At the present time, competition between surrounding rivers to the west and north has stabilized. They all join the Ohio at one point or another and share similar constraints on their long trip to the sea.

The Chattahoochee River (CH.R.), which flows out of the mountains of Georgia, shares the Appalachian trend along part of its course, although the flow is to the southwest directly to the Gulf of Mexico. It, too, is at least 300 million years old. It is no coincidence that these old streams flow to the Gulf of Mexico rather than to the Atlantic Ocean. They are so old that they predate the Atlantic basin, which is just the water-filled hole left by the displacement of the modern continents in the Jurassic period, about 180 million years ago.

Therefore, via least astonishment:

In contrast to the New and the Chattahoochee Rivers, all of the rivers draining into the Atlantic Ocean were born as a result of continental displacement. When the Atlantic Ocean basin began to open, rain would naturally flow off into the new hole. The rivers of the modern Atlantic drainage system certainly flow down short, steep courses to reach sea level, with few constraints to bar their passage. As a result, they have been able to erode headward quite rapidly and have taken over territory that once belonged to older, more constrained tributaries of the Mississippi-Ohio system. The Potomac and James Rivers (P.R., J.R.) have cut well back through the main mass of the mountains and effectively pirated a vast territory that once belonged to the now minor streams of southwestern Pennsylvania and adjacent West Virginia.

Piracy occurs most readily when an active river system approaches an inactive one at right angles and from a lower elevation. Fig. 16 clearly shows that the Chattahoochee and New Rivers (CH.R., N.R.) are about to be attacked by a number of Atlantic streams. These are eroding headward at right angles to the line of the older streams and attacking from points more than 1,000 feet lower. Given a "little" more time, the New River and the Chattahoochee will be turned by piracy into tributaries to the Atlantic drainage. It is all right there on the map before your eyes!

Is it possible that the New River, Tennessee River, and ancient Ohio River system once flowed into the Arctic Ocean? If so, they were captured by the Mississippi River as it extended its drainage basin to the north. That possibility is hard to prove or disprove.

Example, facts 1 to N:

There is a very interesting area in the western United States called the Great Basin (G.B.). All the drainage in this area is internal; no streams reach the sea. The area is bounded on the east by the divide that separates it from the Colorado River drainage (CD.R.), on the north by the Snake River drainage (S.R.), and on the west by the Sierra Nevada (S.N.). This is an arid region containing the Mojave Desert, Death Valley, and Great Salt Lake (S.L.).

The Great Basin has not always been as we see it today. During glacial times the regional climate was wet enough to support two very large lakes in the area—Lahontan and Bonneville. Each, comparable to Lake Michigan in size, was about 1,000 feet deep. Fed by the abundant rain and melted snow, these lakes once overflowed into the Snake River drainage. Now with the return of dry climatic conditions they have evaporated, leaving nothing more than abandoned shorelines, deltas, and salt brine ponds. The Great Salt Lake is one of the more famous lingering brine ponds. There are many other salt flats linked to this story.

Illustrations showing how you use all of these concepts

The present dry climate of the Great Basin is created in part by the high wall of the Sierra Nevada and the Cascade Range. North of the San Francisco area (G.G., Golden Gate), the warm Japan Current in the Pacific Ocean causes such an effective flow of moist air against the western slopes of the mountains that they are covered with rain forests. However, the moist air cannot cross the mountain barrier, and as a result, there are sharply bounded deserts on the eastern side of the mountains.

Soils in arid regions tend to be chemically rich, providing substantial nutrition to plants and the grazing animals that live on them. The explanation lies in the movement of ground water. In dry areas, capillary action brings ground water to the surface between rains. As it evaporates, the water leaves its dissolved salts in the form of a mineral-rich, crusty soil called caliche. Range grasses that grow on caliche do not have the lush appearance of the eastern meadow grasses, but because of their rich mineral content they are better able to support cattle than grass grown in the soils of humid areas, which have been depleted of minerals by the washing effects of the movement of ground water into the earth.

Dry-farming techniques have been practiced on these mineral-rich soils since the earliest days of the agricultural revolution. The great civilizations of Babylon and the Mexican plateau rose as hydraulic civilizations where rich arid land and irrigation became the base for an extensive agricultural economy. Five millenia ago a portion of the population of the arid regions were thus freed for other pursuits in the arts, sciences, and architecture. Rich soils in the dry areas had more to do with the achievements of the Aztecs, Incas, Egyptians, and Babylonians than we normally recognize. The Mormons repeated the achievement when they moved into the Salt Lake area in 1847, as did the Israelis in Israel, particularly after World War II.

The inset map of the western states that is included with Fig. 16 shows the proportion of the region that is still held as federal land. In many ways this is land that no one has wanted, as much of it is too dry by Eastern standards.

Therefore, via least astonishment:
When our population increases to the point at which additional living space becomes a vital necessity, this dry region will have to be developed. All it needs is water in great quantities. Develop-

ment along the Rio Grande and the Columbia and Colorado River systems has been in progress for decades. There are many dams and lakes on these rivers, and the water is used for irrigation to a degree that is approaching the limit of the supply. Yet the Great Basin and much of the surrounding public lands remain undeveloped. A number of plans to obtain water from the north have been proposed.[3-7] All of them would require a great series of dams, artificial river channels, lakes, and pumping stations to bring water from as far away as the Yukon River in Alaska and the Canadian rivers that now flow to the Pacific Ocean or Hudson Bay. Some of the plans would use the Great Lakes as storage reservoirs for water moved to the southwest from the Canadian provinces of Quebec and Manitoba. The most ambitious plans would take water from Alaska, British Columbia, and Alberta as well. The water could be distributed throughout the western states and on into the deserts of northern Mexico in a grand scheme of international cooperation on the most stupendous scale. The Canadians could produce and sell water just as any other product is marketed. Right now the resource is lost to the seas in the northern portion of the North American continent and is woefully lacking in the southern portion.

The empty land that no one wants could be made to bloom through the combined efforts of international statesmen, business enterprises, and the technical expertise of economists, geologists, and engineers. We should recognize that giant cities are unstable because they are too much of a bad thing. One cure for the ills of great cities lies in making alternative places for people to live. We have the capacity to recreate the Great Basin as it was in Pleistocene time, a land of lakes around which new life may flourish. The first Americans who came across the Bering Bridge once camped on shores of lakes that are no longer in existence; think of that!

ANNOTATED REFERENCES

1. Kelley, Fred C. 1951. Miracle at Kitty Hawk. Farrar, Straus & Giroux, Inc., New York. 482 pp. (This statement by Orville Wright appears in a letter dated June 7, 1903, six months and ten days before the first powered flight. It is found on p. 91. Readers are encouraged to examine the genius of the Wright brothers for themselves.)

2. Kestin, Joseph. 1970. Creativity in teaching and learning. Am. Sci. **58**:250-256.

3. Sewell, W. R. Derrick. 1967. NAWAPA: a continental water system, pipedream or practical possibility. Bull. Atom. Sci. **23**:8-13.

4. Ostrom, Vincent. 1967. NAWAPA: a continental water system, political feasibility. Bull. Atom. Sci. **23**:13-16.

5. Crutchfield, James A. 1967. NAWAPA: a continental water system, economic considerations. Bull. Atom. Sci. **23**:17-21.

6. Tinney, E. Roy. 1967. NAWAPA: a continental water system, engineering aspects. Bull. Atom. Sci. **23**:21-25.

7. Royce, William. 1967. NAWAPA: a continental water system, fish and fishing. Bull. Atom. Sci. **23**:26-27.

Epilogue

We have come full cycle. With a new ability "to see a world" we have learned to comprehend the past, to appreciate the present, and to anticipate the future.

Each human life-span remains a typical interval confined between the events of birth and death. Every life-span is subject to the physical limitations imposed by time. Yet the human mind is capable of transcending time, for it may be freed by pure reason and vicarious experiences via imagination. As the poet William Blake knew, an educated mind is the only organ on earth with which one can "Hold Infinity in the palm of your hand/And Eternity in an hour."